Advances in Experimental Medicine and Biology

Volume 933

More information about this series at http://www.springer.com/series/5584

Jong Hoon Park • Curie Ahn
Editors

Cystogenesis

Springer

Editors
Jong Hoon Park
Department of Life systems
Sookmyung Women's University
Seoul, South Korea

Curie Ahn
Department of Internal Medicine
Seoul National University College of Medicine
Seoul, South Korea

ISSN 0065-2598 ISSN 2214-8019 (electronic)
Advances in Experimental Medicine and Biology
ISBN 978-981-10-2040-7 ISBN 978-981-10-2041-4 (eBook)
DOI 10.1007/978-981-10-2041-4

Library of Congress Control Number: 2016955171

Printed on acid-free paper

This Springer imprint is published by Springer Nature
The registered company is Springer Science+Business Media Singapore Pte Ltd.

Preface

Autosomal dominant polycystic kidney disease (ADPKD) is a highly prevalent hereditary renal disorder that affects at least 1 in every 1000 individuals worldwide. The development of numerous fluid-filled cysts is the evident pathology of ADPKD, and it is accompanied by hyperactivation of cell proliferation, interstitial inflammation, and fibrosis, finally reaching end-stage renal disease (ESRD). The age at which ESRD is reached is highly variable in patients with ADPKD, and a vast number of cases show the occurrence of ADPKD later than 50 years of life. In addition, more than 50 % of the patients with ADPKD require dialysis or renal transplantation in their 60s. In fact, ADPKD accounts for up to 10 % of all the renal transplant patients. Though causative genes and major aberrant signaling pathways involved in ADPKD have already been identified, no specific targeted strategies are available to cure ADPKD.

This book entitled "Cystogenesis" comprises a comprehensive review of clinical trials attempted until now and the basic cellular mechanisms of ADPKD. We provide an overview of the molecular mechanisms of ADPKD pathogenesis based on the latest thesis papers. It includes genetic mechanisms, intracellular signaling pathways, and epigenetic regulation. Moreover, other cystogenesis mechanisms stimulated by disrupted primary cilia have also been introduced. A better understanding of such basic mechanisms underlying the onset of ADPKD might provide an important insight to identify the potential therapeutic targets. Furthermore, therapeutic approaches or novel diagnostic biomarkers proposed until now have been reviewed with a mention of their limitations and perspective.

We are very pleased and honored to publish this special issue of Springer on ADPKD. We hope that this book provides a broad overview of ADPKD and deals with the key challenges currently faced by researchers in this field. Furthermore, we wish our review provides great evidence to find novel biomarkers for ADPKD, thereby contributing to the development of therapeutic strategies.

Seoul, South Korea

Jong Hoon Park
Curie Ahn

Contents

Part I ADPKD Overview

1 Recent Trends in ADPKD Research 3
Yu Bin Shin and Jong Hoon Park

2 Genetic Mechanisms of ADPKD 13
Do Yeon Kim and Jong Hoon Park

Part II Cystogenesis Mechanisms

3 Cell Proliferation and Apoptosis in ADPKD 25
Eun Ji Lee

4 Inflammation and Fibrosis in ADPKD 35
Hyowon Mun and Jong Hoon Park

5 Functional Study of the Primary Cilia in ADPKD 45
Je Yeong Ko

6 Epigenetic Regulation in Cystogenesis 59
Yu Mi Woo

Part III Therapeutic Approaches and Diagnostic Markers for ADPKD

7 Validation of Effective Therapeutic Targets for ADPKD Using Animal Models 71
Yu Mi Woo, Je Yeong Ko, and Eun Ji Lee

8 Diagnostic Evaluation as a Biomarker in Patients with ADPKD 85
Hayne Cho Park and Curie Ahn

9 Clinical Trials and a View Toward the Future of ADPKD 105
Hyunsuk Kim and Young-Hwan Hwang

Index 123

Contributors

Curie Ahn Department of Internal Medicine, Seoul National University College of Medicine, Seoul, South Korea

Young-Hwan Hwang Department of Internal Medicine, Eulji General Hospital, Seoul, South Korea

Hyunsuk Kim Department of Internal Medicine, Seoul National University Hospital, Seoul, South Korea

Do Yeon Kim Molecular Medicine Laboratory, Department of Lifc systems, Sookmyung Women's University, Seoul, South Korea

Je Yeong Ko Molecular Medicine Laboratory, Department of Life systems, Sookmyung Women's University, Seoul, South Korea

Eun Ji Lee Molecular Medicine Laboratory, Department of Life systems, Sookmyung Women's University, Seoul, South Korea

Hyowon Mun Molecular Medicine Laboratory, Department of Life systems, Sookmyung Women's University, Seoul, South Korea

Hayne Cho Park Division of Nephrology, Department of Internal Medicine, The Armed Forces Capital Hospital, Seongnam-si, Gyeonggi-do, South Korea

Jong Hoon Park Department of Life systems, Sookmyung Women's University, Seoul, South Korea

Yu Bin Shin Molecular Medicine Laboratory, Department of Life systems, Sookmyung Women's University, Seoul, South Korea

Yu Mi Woo Molecular Medicine Laboratory, Department of Life systems, Sookmyung Women's University, Seoul, South Korea

Part I
ADPKD Overview

Chapter 1
Recent Trends in ADPKD Research

Yu Bin Shin and Jong Hoon Park

Abstract Autosomal Dominant Polycystic Kidney Disease (ADPKD) is one of the most common inherited disorders. It is the fourth leading cause of renal replacement and renal failure worldwide. Mutations in *PKD1* or *PKD2* cause ADPKD. Patients with ADPKD show progressive growth of renal cysts filled with cystic fluid, leading to end-stage renal disease (ESRD) and renal failure by their sixth decade of life. Currently, there are no curative treatments for ADPKD. Therefore, patients require dialysis or kidney transplantation. To date, researchers have elucidated many of the mechanisms that cause ADPKD and developed many methods to diagnose the disease. ADPKD is related to growth factors, signaling pathways, cell proliferation, apoptosis, inflammation, the immune system, structural abnormalities, epigenetic mechanisms, microRNAs, and so on. Various therapies have been reported to slow the progression of ADPKD and alleviate its symptoms.

Keywords ADPKD • Polycystic kidney • Cyst • Renal failure • ESRD • Pathogenesis • Disease mechanism

1.1 Autosomal Dominant Polycystic Kidney Disease

1.1.1 Pathogenesis of ADPKD

Three inherited cystic diseases of the kidney are known. These are autosomal dominant polycystic liver disease (ADPLD), autosomal recessive polycystic kidney disease (ARPKD), and autosomal dominant polycystic kidney disease (ADPKD). *PRKCSH* and *SEC63* genes are involved in ADPLD, and it causes bile duct cystic

Y.B. Shin
Molecular Medicine Laboratory, Department of Life systems, Sookmyung Women's University, Cheongpa-ro 47-gil 100, Yongsan-gu, Seoul 04310, South Korea
e-mail: binnie90@sookmyung.ac.kr

J.H. Park (✉)
Department of Life systems, Sookmyung Women's University, Cheongpa-ro 47-gil 100, Yongsan-gu, Seoul 04310, South Korea
e-mail: parkjh@sookmyung.ac.kr

J.H. Park, C. Ahn (eds.), *Cystogenesis*, Advances in Experimental Medicine and Biology 933, DOI 10.1007/978-981-10-2041-4_1

dilations. Abnormal expression of the *PKHD1* gene results in ARPKD, which manifests as fusiform collecting duct dilatations, congenital hepatic fibrosis, and liver cysts.

Polycystic kidney disease 1 and 2 (*PKD1* and *PKD2*) are two causative genes of ADPKD. Patients with ADPKD have kidney and bile duct cysts. Germline mutation of *PKD1* or *PKD2* leads to cyst formation. Monogenic disorders of these two genes have an incidence of 1 in 600 to 1 in 1000 individuals, meaning that ADPKD is one of the most common hereditary disorders. Also, it is the fourth leading cause of renal replacement and renal failure worldwide (Fedeles et al. 2014). Most cases of ADPKD are inherited, while about 10 % are due to *de novo* mutation (Rossetti et al. 2001).

1.1.2 Manifestations of ADPKD

Patients with ADPKD have symptoms as follows: They have fluid-filled renal cysts in their kidneys and other epithelial organs. Usually, both of their kidneys are extremely enlarged and filled with cystic fluid. Because of the formation of cysts, patients have renal enlargement and it eventually causes end-stage renal failure (ESRD) (Hou et al. 2002). Approximately 10 % of ESRD cases are caused by ADPKD. In normal adult humans, the kidneys comprise about 0.5 % of body weight. The kidneys of ADPKD patients weigh about 22 kg, comprising about 20 % of body weight (Ekser and Rigotti 2010). There are also various manifestations of ADPKD that are unrelated to the kidneys. One of the most common manifestations in ADPKD patients is hypertension (Martinez and Grantham 1995). Connective tissue defects such as cardiac defects, intracranial aneurysms, and hepatocystic disease also occur in ADPKD patients (Wu et al. 2000; Hughes et al. 1995). When these symptoms become aggravated, patients have to undergo dialysis and transplantation.

One of the most pronounced phenotypes of ADPKD is cyst formation in the kidneys. When the disease becomes severe, patients with enlarged cysts can even resemble pregnant women. Then, how does cyst formation progress in the kidney?

First, it starts at the normal renal tubule. Germline mutation of *PKD1* or *PKD2* causes the loss of one allele, and a somatic second hit causes the loss of the other normal allele (explained in Chap. 2). Then, one or more additional steps lead to cystogenesis as a ‘third hit’, which may include nephrotoxic injury and/or ischemia. The third hit leads to cell proliferation in the renal tubules and causes small dilations that subsequently expand and form fluid-filled cysts of various sizes (Weimbs 2007; Bell et al. 2011). As cell proliferation persists, the dilation also progresses. Then, the dilated regions are separated from their original tubules, becoming a distinct cyst (Weimbs 2011). Once cysts are formed, they grow increasingly larger as the disease progresses. Therefore, cystogenesis is the most remarkable feature of ADPKD.

In addition, a variety of mechanisms can evoke ADPKD disease progression and cystogenesis, including somatic mutation, germ line mutation, modifying genes, increased cell proliferation, apoptosis, defective planar cell polarity, extracellular matrix abnormalities, fluid secretion, inflammation, and environmental factors (Paul and Vanden Heuvel 2014; Zhou 2009). To date, researchers have elucidated a variety of mechanisms that can cause ADPKD, methods to diagnose the disease, and therapies to slow and alleviate the symptoms of ADPKD.

1.2 Studies on ADPKD

As many studies on other diseases have investigated their early diagnosis and treatments, those on ADPKD have also sought to identify the mechanisms of disease and to develop methods to cure ADPKD. For several decades, researchers have been working towards these goals. Some of most remarkable approaches that have been employed to study ADPKD are as follows.

1.2.1 Growth Factors, Signaling Pathways, and Cell Proliferation

Cyst expansion in the kidneys of ADPKD patients is the most remarkable feature of the disease. The tubular epithelial cells that surround the cysts proliferate and drive their enlargement. Therefore, inhibiting cell proliferation is an important target for easing the symptoms of ADPKD (LaRiviere et al. 2015). In normal kidneys, a homeostatic balance is maintained between cell proliferation and apoptosis. However, in the kidneys of ADPKD patients, cell proliferation is more frequent than apoptosis. This imbalance eventually leads to cyst formation (Gregoire et al. 1987). Many studies have suggested a variety of mechanisms that may cause or contribute to cell proliferation and cyst enlargement in cystic kidney epithelia.

A wide range of growth factors are involved in cystogenesis. One of the main growth factors is epidermal growth factor (EGF), which along with EGF receptor, and other members of the EGF family such as transforming growth factor (TGF)-α, heparin-binding EGT, and amphiregulin, plays an important role in regulating cell proliferation of the cystic epithelia (Du and Wilson 1995). TGF-α is overexpressed in human polycystic kidney. Also, TGF-β is a major growth factor in ADPKD. Upregulation of TGF-β is related to cyst expansion during disease progression, but it is less involved in cyst initiation (Hassane et al. 2010; Wilson et al. 1996). Other growth factors such as hepatocyte growth factor (HGF), insulin-like growth factor 1 (IGF-1), and tyrosine kinase receptor of HGF and IGF-1 are also related to cyst formation in ADPKD.

In addition to growth factors, many signaling pathways are involved in ADPKD. First of all, the second messenger adenosine 3′, 5′ cyclic monophosphate (cAMP) is crucial in cystic kidney. cAMP, an intracellular mediator of adenylyl cyclase signaling, promotes cell proliferation of cystic epithelia. ADPKD patients and various animal models of polycystic kidney disease show an elevated cAMP level in kidney. Even in normal human kidney, stimulation by cAMP drives a cystic phenotype. Upregulation of cAMP mainly influences calcium signaling. cAMP induces cyst expansion and fluid secretion in intact cysts (Ye and Grantham 1993). When the level of intracellular calcium is decreased, cAMP/PKA signaling activates the Src/Ras/Raf/MEK/ERK pathways in ADPKD patients. ERK signaling, which is induced by PKA, also leads to the activation of mTOR signaling (Spirli et al. 2010; Distefano et al. 2009). Signal transducer and activator of transcription 3 (STAT3) responses to cAMP. STAT3 plays key roles in the development and maintenance of proinflammatory conditions in cystic kidneys (Martinez and Grantham 1995; Talbot et al. 2014).

Following the identification of clear mechanisms driving cell proliferation in the cysts of ADPKD, there have been many clinical trials to alleviate disease progression. Some representative drugs are vasopressin V2 receptor (V2R) antagonist, somatostatin analogs, mTOR inhibitors (rapamycin, sirolimus, and everolimus), Raf kinase inhibitors (PLX5568 and sorafenib), Src/Abl inhibitor SKI-606 (bosutinib), and MEK inhibitors (PD184653 and UO126) (Renken et al. 2011; Sweeney et al. 2008; Shillingford et al. 2006; Elliott et al. 2011; Omori et al. 2006; Tao et al. 2005; Buchholz et al. 2011; Shibazaki et al. 2008; Yamaguchi et al. 2010).

1.2.2 Inflammation and Immune System

Recently, activation of macrophages was detected in the cyst lining epithelia in several mouse models of ADPKD. This find means that activated macrophages are involved in the proliferation of tubular epithelial cells (Karihaloo et al. 2011; Rae et al. 2007; Swenson-Fields et al. 2013). In addition, some inflammatory responses and gene expression patterns associated with the immune system were determined to be involved in ADPKD through computational analysis (Song et al. 2009). An increased concentration of monocyte chemotactic protein-1 (MCP-1) induces an increased number of mononuclear cells. We can detect MCP-1 in urine samples from ADPKD patients, which indicates an impairment of the innate immune system because of disease progression (Swenson-Fields et al. 2013). Macrophages play a major role in early developmental stages by removing apoptotic cells after the differentiation of organs. An increased number of macrophages results in the up-regulation of cell proliferation, down-regulation of apoptosis, and finally enlargement of cysts in the kidneys. Several studies have suggested that the downregulation of macrophages might be helpful for curing polycystic kidney diseases.

1.2.3 Structural Abnormality

The progressive accumulation of extracellular matrix is one of the notable hallmarks of fibrosis in ADPKD. Some animal models of polycystic kidney disease exhibit a thickened and laminated basement membrane and express high levels of α1 type IV collagen and laminins β1 and β2 (Katz et al. 1989). Polycystin-1, the protein encoded by *PKD1*, is involved in interactions between cells and the extracellular matrix. An excessive accumulation of fibroblasts results in cyst expansion in ADPKD kidney. Recently, in studies using zebrafish, researchers have found that polycystin proteins might be engaged in producing collagen. Therefore, we can infer that the accumulation of collagen is due to malfunctions of *PKD1* and *PKD2* (Mangos et al. 2010).

Extracellular matrix maintains its structure through continual turnover. The rate of the degradation of extracellular matrix is mediated by the matrix metalloproteinases (MMPs), and tissue inhibitors of metalloproteinases (TIMPs) (Catania et al. 2007). In the kidneys of a mouse model with a mutation in *Pkd1*, the levels of MMP-2 and MMP-14 were upregulated (Hassane et al. 2010). Moreover, the levels of MMP-1, MMP-9, and TIMP-1 in serum were increased in the kidneys of ADPKD patients (Nakamura et al. 2000). Taken together, most MMPs and TIMPs were elevated in cystic conditions. In addition, the extracellular matrix interacts with cells. This interaction controls cell proliferation, differentiation, and other cellular functions through specific matrix receptor proteins. Typical examples of these matrix receptor proteins are integrins and proteoglycan-containing syndecans (Geiger et al. 2001). To summarize many studies, researchers found that several integrins and syndecans were increased in ADPKD patients, particularly α2β1 integrin, integrin α8, integrin αv, integrin β4, and syndecan-4 (Wilson and Burrow 1999; Wallace et al. 2008; Wilson et al. 1999; Zeltner et al. 2008).

The main physiological function of the kidney is filtration, and it is essential for homeostasis. Components that are over-accumulated or unnecessary are secreted, while other essential factors remain in the circulation. This process occurs when body fluid passes through the kidney, especially in the renal tubules. A sensory organelle called the cilium can detect physical and chemical stimuli such as the flow of fluid (Hildebrandt and Otto 2005). Cilia protrude from the epithelial cells towards the lumen of renal tubules. Cilia are microtubule-based structures and they originate from the basal body or centrosome (Paul and Vanden Heuvel 2014). Alongside many other factors, ciliary defects can also cause polycystic kidney disease. Recent studies have demonstrated that malfunctions of the cilia could influence cyst development (Garcia-Gonzalo and Reiter 2012).

1.2.4 Epigenetic Changes and microRNA

Another biological mechanism that could explain the symptoms and pathogenesis of ADPKD is epigenetic regulation. Epigenetic changes including histone modifications such as acetylation, methylation, and phosphorylation are also related to the mechanisms of ADPKD (Li 2011). The mechanism that can control DNA methylation was revealed—an increased level of TGF-β evokes DNA methylation in ADPKD tissue, and it might result in fibrosis in kidney (Bechtel et al. 2010).

Furthermore, microRNAs can also evoke ADPKD by directly binding to their target genes. The importance of microRNAs in ADPKD has been emphasized. microRNAs can regulate the expression level of their target mRNA(s) and are related to cell proliferation, differentiation, apoptosis, and many other cellular processes. Individual microRNAs might be increased or decreased in the kidneys of ADPKD patients. Some microRNAs target *PKD1* or *PKD2* directly, or they can target other genes related to the phenotype of ADPKD. For example, the miR-17 ~ 92 microRNA cluster, miRNA-21, miR-15a, and miRNA-199a have been identified as candidate ADPKD-involved microRNAs that can regulate cell proliferation and the pathogenesis of ADPKD (Patel et al. 2013; Sun et al. 2015; Lakhia et al. 2015; Lee et al. 2008).

Targeting microRNAs that bind to *PKD1* or *PKD2* might be a viable method to regulate the clinical manifestations of ADPKD, but most previous research about microRNAs and ADPKD has suggested that microRNAs alone are not sufficient for the treatment of ADPKD. Another factor is needed to cure the disease, so many studies have offered microRNAs and their target genes as new therapeutic targets. As a variety of microRNAs involved in cystogenesis have begun to be revealed by microRNA microarray data, research on microRNAs in ADPKD is predicted to become increasingly active in the future (Tan et al. 2011).

1.3 What Is Coming Next in ADPKD Research?

Until now, research on ADPKD has been conducted using various approaches such as those at the molecular and structural levels, as well as clinical trials. Although all of these approaches can yield useful results, we need to increasingly focus on identifying methods for the diagnosis and treatment of ADPKD. All researchers should seek to develop a framework for integrating studies across disciplines, and then apply that framework to deciphering an effective treatment modality for curing ADPKD. As many exciting studies are currently underway, it is possible that we may be able to discover a new therapeutic method in the near future.

References

Bechtel W, McGoohan S, Zeisberg EM, Muller GA, Kalbacher H, Salant DJ, Muller CA, Kalluri R, Zeisberg M (2010) Methylation determines fibroblast activation and fibrogenesis in the kidney. Nat Med 16(5):544–550. doi:10.1038/nm.2135

Bell PD, Fitzgibbon W, Sas K, Stenbit AE, Amria M, Houston A, Reichert R, Gilley S, Siegal GP, Bissler J, Bilgen M, Chou PC, Guay-Woodford L, Yoder B, Haycraft CJ, Siroky B (2011) Loss of primary cilia upregulates renal hypertrophic signaling and promotes cystogenesis. J Am Soc Nephrol 22(5):839–848. doi:10.1681/ASN.2010050526

Buchholz B, Klanke B, Schley G, Bollag G, Tsai J, Kroening S, Yoshihara D, Wallace DP, Kraenzlin B, Gretz N, Hirth P, Eckardt KU, Bernhardt WM (2011) The Raf kinase inhibitor

PLX5568 slows cyst proliferation in rat polycystic kidney disease but promotes renal and hepatic fibrosis. Nephrol Dial Transplant 26(11):3458–3465. doi:10.1093/ndt/gfr432
Catania JM, Chen G, Parrish AR (2007) Role of matrix metalloproteinases in renal pathophysiologies. Am J Physiol Renal Physiol 292(3):F905–F911. doi:10.1152/ajprenal.00421.2006
Distefano G, Boca M, Rowe I, Wodarczyk C, Ma L, Piontek KB, Germino GG, Pandolfi PP, Boletta A (2009) Polycystin-1 regulates extracellular signal-regulated kinase-dependent phosphorylation of tuberin to control cell size through mTOR and its downstream effectors S6K and 4EBP1. Mol Cell Biol 29(9):2359–2371. doi:10.1128/MCB.01259-08
Du J, Wilson PD (1995) Abnormal polarization of EGF receptors and autocrine stimulation of cyst epithelial growth in human ADPKD. Am J Physiol 269(2 Pt 1):C487–C495
Ekser B, Rigotti P (2010) Images in clinical medicine. Autosomal dominant polycystic kidney disease. N Engl J Med 363(1):71. doi:10.1056/NEJMicm0905399
Elliott J, Zheleznova NN, Wilson PD (2011) c-Src inactivation reduces renal epithelial cell-matrix adhesion, proliferation, and cyst formation. Am J Physiol Cell Physiol 301(2):C522–C529. doi:10.1152/ajpcell.00163.2010
Fedeles SV, Gallagher AR, Somlo S (2014) Polycystin-1: a master regulator of intersecting cystic pathways. Trends Mol Med 20(5):251–260. doi:10.1016/j.molmed.2014.01.004
Garcia-Gonzalo FR, Reiter JF (2012) Scoring a backstage pass: mechanisms of ciliogenesis and ciliary access. J Cell Biol 197(6):697–709. doi:10.1083/jcb.201111146
Geiger B, Bershadsky A, Pankov R, Yamada KM (2001) Transmembrane crosstalk between the extracellular matrix--cytoskeleton crosstalk. Nat Rev Mol Cell Biol 2(11):793–805. doi:10.1038/35099066
Gregoire JR, Torres VE, Holley KE, Farrow GM (1987) Renal epithelial hyperplastic and neoplastic proliferation in autosomal dominant polycystic kidney disease. Am J Kidney Dis 9(1):27–38
Hassane S, Leonhard WN, van der Wal A, Hawinkels LJ, Lantinga-van Leeuwen IS, ten Dijke P, Breuning MH, de Heer E, Peters DJ (2010) Elevated TGFbeta-Smad signalling in experimental Pkd1 models and human patients with polycystic kidney disease. J Pathol 222(1):21–31. doi:10.1002/path.2734
Hildebrandt F, Otto E (2005) Cilia and centrosomes: a unifying pathogenic concept for cystic kidney disease? Nat Rev Genet 6(12):928–940. doi:10.1038/nrg1727
Hou X, Mrug M, Yoder BK, Lefkowitz EJ, Kremmidiotis G, D'Eustachio P, Beier DR, Guay-Woodford LM (2002) Cystin, a novel cilia-associated protein, is disrupted in the cpk mouse model of polycystic kidney disease. J Clin Invest 109(4):533–540. doi:10.1172/JCI14099
Hughes J, Ward CJ, Peral B, Aspinwall R, Clark K, San Millan JL, Gamble V, Harris PC (1995) The polycystic kidney disease 1 (PKD1) gene encodes a novel protein with multiple cell recognition domains. Nat Genet 10(2):151–160. doi:10.1038/ng0695-151
Karihaloo A, Koraishy F, Huen SC, Lee Y, Merrick D, Caplan MJ, Somlo S, Cantley LG (2011) Macrophages promote cyst growth in polycystic kidney disease. J Am Soc Nephrol 22(10):1809–1814. doi:10.1681/ASN.2011010084
Katz SK, Hakki A, Miller AS, Finkelstein SD (1989) Ultrastructural tubular basement membrane lesions in adult polycystic kidney disease. Ann Clin Lab Sci 19(5):352–359
Lakhia R, Hajarnis S, Williams D, Aboudehen K, Yheskel M, Xing C, Hatley ME, Torres VE, Wallace DP, Patel V (2015) MicroRNA-21 aggravates cyst growth in a model of polycystic kidney disease. J Am Soc Nephrol. doi:10.1681/ASN.2015060634
LaRiviere WB, Irazabal MV, Torres VE (2015) Novel therapeutic approaches to autosomal dominant polycystic kidney disease. Transl Res 165(4):488–498. doi:10.1016/j.trsl.2014.11.003
Lee SO, Masyuk T, Splinter P, Banales JM, Masyuk A, Stroope A, Larusso N (2008) MicroRNA15a modulates expression of the cell-cycle regulator Cdc25A and affects hepatic cystogenesis in a rat model of polycystic kidney disease. J Clin Invest 118(11):3714–3724. doi:10.1172/JCI34922
Li X (2011) Epigenetics and autosomal dominant polycystic kidney disease. Biochim Biophys Acta 1812(10):1213–1218. doi:10.1016/j.bbadis.2010.10.008

Mangos S, Lam PY, Zhao A, Liu Y, Mudumana S, Vasilyev A, Liu A, Drummond IA (2010) The ADPKD genes pkd1a/b and pkd2 regulate extracellular matrix formation. Dis Model Mech 3(5–6):354–365. doi:10.1242/dmm.003194

Martinez JR, Grantham JJ (1995) Polycystic kidney disease: etiology, pathogenesis, and treatment. Dis Mon 41(11):693–765

Nakamura T, Ushiyama C, Suzuki S, Ebihara I, Shimada N, Koide H (2000) Elevation of serum levels of metalloproteinase-1, tissue inhibitor of metalloproteinase-1 and type IV collagen, and plasma levels of metalloproteinase-9 in polycystic kidney disease. Am J Nephrol 20(1):32–36. doi:13552

Omori S, Hida M, Fujita H, Takahashi H, Tanimura S, Kohno M, Awazu M (2006) Extracellular signal-regulated kinase inhibition slows disease progression in mice with polycystic kidney disease. J Am Soc Nephrol 17(6):1604–1614. doi:10.1681/ASN.2004090800

Patel V, Williams D, Hajarnis S, Hunter R, Pontoglio M, Somlo S, Igarashi P (2013) miR-17 ~ 92 miRNA cluster promotes kidney cyst growth in polycystic kidney disease. Proc Natl Acad Sci U S A 110(26):10765–10770. doi:10.1073/pnas.1301693110

Paul BM, Vanden Heuvel GB (2014) Kidney: polycystic kidney disease. Wiley Interdiscip Rev Dev Biol 3(6):465–487. doi:10.1002/wdev.152

Rae F, Woods K, Sasmono T, Campanale N, Taylor D, Ovchinnikov DA, Grimmond SM, Hume DA, Ricardo SD, Little MH (2007) Characterisation and trophic functions of murine embryonic macrophages based upon the use of a Csf1r-EGFP transgene reporter. Dev Biol 308(1):232–246. doi:10.1016/j.ydbio.2007.05.027

Renken C, Fischer DC, Kundt G, Gretz N, Haffner D (2011) Inhibition of mTOR with sirolimus does not attenuate progression of liver and kidney disease in PCK rats. Nephrol Dial Transplant 26(1):92–100. doi:10.1093/ndt/gfq384

Rossetti S, Strmecki L, Gamble V, Burton S, Sneddon V, Peral B, Roy S, Bakkaloglu A, Komel R, Winearls CG, Harris PC (2001) Mutation analysis of the entire PKD1 gene: genetic and diagnostic implications. Am J Hum Genet 68(1):46–63. doi:10.1086/316939

Shibazaki S, Yu Z, Nishio S, Tian X, Thomson RB, Mitobe M, Louvi A, Velazquez H, Ishibe S, Cantley LG, Igarashi P, Somlo S (2008) Cyst formation and activation of the extracellular regulated kinase pathway after kidney specific inactivation of Pkd1. Hum Mol Genet 17(11):1505–1516. doi:10.1093/hmg/ddn039

Shillingford JM, Murcia NS, Larson CH, Low SH, Hedgepeth R, Brown N, Flask CA, Novick AC, Goldfarb DA, Kramer-Zucker A, Walz G, Piontek KB, Germino GG, Weimbs T (2006) The mTOR pathway is regulated by polycystin-1, and its inhibition reverses renal cystogenesis in polycystic kidney disease. Proc Natl Acad Sci U S A 103(14):5466–5471. doi:10.1073/pnas.0509694103

Song X, Di Giovanni V, He N, Wang K, Ingram A, Rosenblum ND, Pei Y (2009) Systems biology of autosomal dominant polycystic kidney disease (ADPKD): computational identification of gene expression pathways and integrated regulatory networks. Hum Mol Genet 18(13):2328–2343. doi:10.1093/hmg/ddp165

Spirli C, Okolicsanyi S, Fiorotto R, Fabris L, Cadamuro M, Lecchi S, Tian X, Somlo S, Strazzabosco M (2010) ERK1/2-dependent vascular endothelial growth factor signaling sustains cyst growth in polycystin-2 defective mice. Gastroenterology 138(1):360–371. e367. 10.1053/j.gastro.2009.09.005

Sun L, Zhu J, Wu M, Sun H, Zhou C, Fu L, Xu C, Mei C (2015) Inhibition of MiR-199a-5p reduced cell proliferation in autosomal dominant polycystic kidney disease through targeting CDKN1C. Med Sci Monit 21:195–200. doi:10.12659/MSM.892141

Sweeney WE Jr, von Vigier RO, Frost P, Avner ED (2008) Src inhibition ameliorates polycystic kidney disease. J Am Soc Nephrol 19(7):1331–1341. doi:10.1681/ASN.2007060665

Swenson-Fields KI, Vivian CJ, Salah SM, Peda JD, Davis BM, van Rooijen N, Wallace DP, Fields TA (2013) Macrophages promote polycystic kidney disease progression. Kidney Int 83(5):855–864. doi:10.1038/ki.2012.446

Talbot JJ, Song X, Wang X, Rinschen MM, Doerr N, LaRiviere WB, Schermer B, Pei YP, Torres VE, Weimbs T (2014) The cleaved cytoplasmic tail of polycystin-1 regulates Src-dependent STAT3 activation. J Am Soc Nephrol 25(8):1737–1748. doi:10.1681/ASN.2013091026

Tan YC, Blumenfeld J, Rennert H (2011) Autosomal dominant polycystic kidney disease: genetics, mutations and microRNAs. Biochim Biophys Acta 1812(10):1202–1212. doi:10.1016/j.bbadis.2011.03.002

Tao Y, Kim J, Schrier RW, Edelstein CL (2005) Rapamycin markedly slows disease progression in a rat model of polycystic kidney disease. J Am Soc Nephrol 16(1):46–51. doi:10.1681/ASN.2004080660

Wallace DP, Quante MT, Reif GA, Nivens E, Ahmed F, Hempson SJ, Blanco G, Yamaguchi T (2008) Periostin induces proliferation of human autosomal dominant polycystic kidney cells through alphaV-integrin receptor. Am J Physiol Renal Physiol 295(5):F1463–F1471. doi:10.1152/ajprenal.90266.2008

Weimbs T (2007) Polycystic kidney disease and renal injury repair: common pathways, fluid flow, and the function of polycystin-1. Am J Physiol Renal Physiol 293(5):F1423–F1432. doi:10.1152/ajprenal.00275.2007

Weimbs T (2011) Third-hit signaling in renal cyst formation. J Am Soc Nephrol 22(5):793–795. doi:10.1681/ASN.2011030284

Wilson PD, Burrow CR (1999) Cystic diseases of the kidney: role of adhesion molecules in normal and abnormal tubulogenesis. Exp Nephrol 7(2):114–124. doi:20592

Wilson PD, Norman JT, Kuo NT, Burrow CR (1996) Abnormalities in extracellular matrix regulation in autosomal dominant polycystic kidney disease. Contrib Nephrol 118:126–134

Wilson PD, Geng L, Li X, Burrow CR (1999) The PKD1 gene product, "polycystin-1," is a tyrosine-phosphorylated protein that colocalizes with alpha2beta1-integrin in focal clusters in adherent renal epithelia. Lab Invest 79(10):1311–1323

Wu G, Markowitz GS, Li L, D'Agati VD, Factor SM, Geng L, Tibara S, Tuchman J, Cai Y, Park JH, van Adelsberg J, Hou H Jr, Kucherlapati R, Edelmann W, Somlo S (2000) Cardiac defects and renal failure in mice with targeted mutations in Pkd2. Nat Genet 24(1):75–78. doi:10.1038/71724

Yamaguchi T, Reif GA, Calvet JP, Wallace DP (2010) Sorafenib inhibits cAMP-dependent ERK activation, cell proliferation, and in vitro cyst growth of human ADPKD cyst epithelial cells. Am J Physiol Renal Physiol 299(5):F944–F951. doi:10.1152/ajprenal.00387.2010

Ye M, Grantham JJ (1993) The secretion of fluid by renal cysts from patients with autosomal dominant polycystic kidney disease. N Engl J Med 329(5):310–313. doi:10.1056/NEJM199307293290503

Zeltner R, Hilgers KF, Schmieder RE, Porst M, Schulze BD, Hartner A (2008) A promoter polymorphism of the alpha 8 integrin gene and the progression of autosomal-dominant polycystic kidney disease. Nephron Clin Pract 108(3):c169–c175. doi:10.1159/000116887

Zhou J (2009) Polycystins and primary cilia: primers for cell cycle progression. Annu Rev Physiol 71:83–113. doi:10.1146/annurev.physiol.70.113006.100621

Chapter 2
Genetic Mechanisms of ADPKD

Do Yeon Kim and Jong Hoon Park

Abstract Autosomal dominant polycystic kidney disease is caused by mutation of PKD1 (polycystic kidney disease-1) or PKD2 (polycystic kidney disease-2). PKD1 and PKD2 encode PC1 (polycystin-1) and PC2 (polycystin-2), respectively. In addition, the mutation of cilia-associated proteins is also a recognized major factor of pathogenesis, since PC1 and PC2 are located in primary cilium. Abnormalities of PC1 or PC2 lead to aberrant signaling through downstream pathways, such as the negative growth regulation, G protein activation, and canonical and non-canonical Wnt pathways. According to the "second hit" model, an additional somatic mutation results in the expansion of cyst growth. In this chapter we discuss the genetic mechanisms and signaling pathways involved in ADPKD.

Keywords Genetic mechanism • PKD • Mutation • PKD1 • PKD2 • Polycystin-1 • Polycystin-2 • Signaling pathways

2.1 Polycystin-1 and Polycystin-2

Autosomal dominant polycystic kidney disease (ADPKD) is caused by mutations of two genes, namely PKD1 (polycystic kidney disease-1) and PKD2 (polycystic kidney disease-2), which encode polycystin-1 (PC1) and polycystin-2 (PC2), respectively. Approximately 85 % of ADPKD cases result from mutations in PKD1. PC1 is a 450-kD receptor-like protein with a large extracellular N terminus, 11 membrane-spanning domains, and a short cytoplasmic C terminus. The expression of PC1 was evaluated in epithelial cells during development (Ward et al. 1996), but its expression level is high in fetal renal tissue only and low in adult tissue (Chauvet et al.

D.Y. Kim
Molecular Medicine Laboratory, Department of Life systems, Sookmyung Women's University, Cheongpa-ro 47-gil 100, Youngsan-gu, Seoul 04310, South Korea
e-mail: doyeon42@sookmyung.ac.kr

J.H. Park (✉)
Department of Life systems, Sookmyung Women's University, Cheongpa-ro 47-gil 100, Youngsan-gu, Seoul 04310, South Korea
e-mail: parkjh@sookmyung.ac.kr

J.H. Park, C. Ahn (eds.), *Cystogenesis*, Advances in Experimental Medicine and Biology 933, DOI 10.1007/978-981-10-2041-4_2

2002). PC1 localizes to the cilium, plasma membrane, and adhesion complex in polarized epithelial cells (Ibraghimov-Beskrovnaya et al. 1997; Huan and van Adelsberg 1999). PC1 and PC2 form a complex through their C-terminal tails (CTTs) to play a role in intracellular Ca^{2+} regulation (Tsiokas et al. 1997; Qian et al. 1997). The N terminus of PC1 consists of 15 PKD repeat motifs, two leucine-rich motifs, and a C-type lectin domain (Harris et al. 1995; Bycroft et al. 1999). These domains play an important role in mediating the subcellular localization of PC1 within the plasma membrane and junctional complexes (Streets et al. 2009; Babich et al. 2004). The CTT of PC1 includes a G protein-binding domain, a coiled-coil domain, and residues associated with ubiquitin-mediated degradation (Low et al. 2006). The N- and C-terminal domains of PC1 can be cleaved. The N-terminal domain is cleaved at the G protein-coupled receptor proteolysis site (GPS) in the early secretory pathway (Wei et al. 2007; Yu et al. 2007). Generally, PC1 exists as a heterogeneous population of the full-length and N-terminal cleaved forms (Wei et al. 2007). However, one study suggested that N-terminal cleavage is necessary for the complete functional activation of PC1 (Qian et al. 2002). One interesting study revealed that cleaved CTT, which is assembled in the nucleus during reduced fluid flow in mouse kidney, was increased in the cyst-lining cells in ADPKD (Low et al. 2006).

PC2 (968 aa; ~110 kDa) is a six-transmembrane protein with intracellular N and C termini (Mochizuki et al. 1996). PC2 acts as a Ca^{2+}–responsive cation channel of the transient receptor potential family (Gonzalez-Perrett et al. 2001). Although PC2 is co-localized with PC1 to the cilium and plasma membrane (Yoder et al. 2002; Yu et al. 2009), the major portion of cellular PC2 is observed in the intracellular compartment and functions to release calcium from the intracellular store (Vassilev et al. 2001). The channel formed by PC1 and PC2 in complex is activated in response to ciliary bending, and it leads to signal transduction by chemical or mechanical stimuli (Nauli et al. 2003). The calcium-conducting pore consists of the loop between the fifth and sixth transmembrane domains, and a missense mutation in the conducting pore was shown to cause ADPKD (Koulen et al. 2002). PC2 also functions as an indirect regulator of the cytoplasmic calcium level together with two other intracellular Ca^{2+} channels, namely the inositol 1, 4, 5-triphosphate receptor (IP3R) and ryanodine receptor. The C-terminus of PC2, which directly interacts with IP3R, results in IP3-induced Ca^{2+} flux. PC2 also binds to the ryanodine receptor channel and regulates calcium-induced calcium release (Anyatonwu et al. 2007; Li et al. 2009). The largest pools of PC2 appear in the ER and the early Golgi body among the subcellular compartments (Cai et al. 1999; Koulen et al. 2002). PC2's subcellular localization requires specific signal transduction and trafficking proteins that bind to PC2's C terminus (Chapin and Caplan 2010). The movement of PC2 from the ER to the Golgi is modulated by polycystin-2 interactor (PIGEA-14), which causes a redistribution of PC2 (Hidaka et al. 2004).

The PC1 and PC2 proteins are co-located in the primary cilium and ER (Yoder et al. 2002); however, they are also found in other locations depending on their functions (Hanaoka et al. 2000; Grimm et al. 2003). Especially, many studies have suggested that PC1 and PC2 reciprocally influence each other's localization. One study confirmed that impairing the function of PC1 prevents GPS cleavage in ADPKD cyst cells, which leads to a decreased co-localization and amount of both PC1 and PC2 in primary cilia (Xu et al. 2007). The interaction between PC1 and PC2 has been recognized as an important factor for creating a functional ion channel through the intrinsic channel formed by activated PC2 alone or the emergent channel formed by the PC1-PC2 complex (Hanaoka et al. 2000). A physical connection between PC1 and PC2 is mediated by the CTTs of PC1 and PC2 (Qian et al. 1997; Tsiokas et al. 1997; Casuscelli et al. 2009). Through this interaction, PC2 prevents the ability of PC1 to activate G proteins (Delmas et al. 2002).

2.2 Signaling Pathways of PKD1 and PKD2

Although the complete pathologic mechanisms remain to be elucidated, the loss of function of the PC1 and/or PC2 proteins leads to ADPKD pathogenesis through a myriad of signaling pathways, including planar cell polarity (PCP), Wnt, mammalian target of rapamycin (mTOR), cyclic adenosine monophosphate (cAMP), G-protein coupled receptor (GPCR), cystic fibrosis transmembrane conductance regulator (CFTR), epidermal growth factor receptor (EGFR), mitogen-activated protein kinase (MAPK), cellular Ca^{2+}, and the cell cycle (Gallagher et al. 2010). As explained above, PC1 and PC2 form a complex that functions as a transient receptor potential channel to maintain intracellular calcium homeostasis (Vassilev et al. 2001; Anyatonwu and Ehrlich 2005) and as a calcium release channel (Koulen et al. 2002). Disruption of PC1/PC2 results in a decreased level of intracellular Ca^{2+}, leading to upregulated cAMP signaling and increased cell proliferation (Masyuk et al. 2006) (Fig. 2.1).

An increased level of cAMP has been identified in many animal models of polycystic kidney disease (PKD), not only in the kidney but also in other tissues, such as

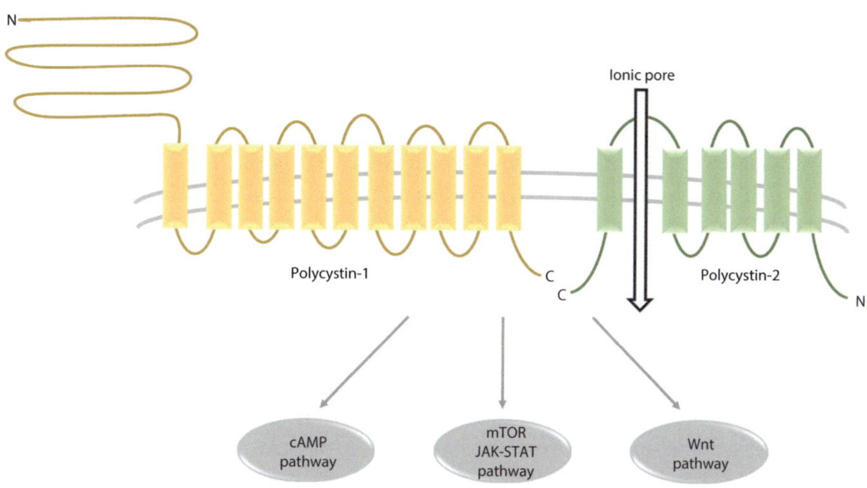

Fig. 2.1 Representative signaling pathways regulated by polycystin 1 and/or 2

cholangiocytes (Masyuk et al. 2007), vascular smooth muscle cells (Kip et al. 2005), and the choroid plexus (Banizs et al. 2007). cAMP levels are modulated by the activities of membrane-bound GPCRs, soluble adenylyl cyclases (ACs), and cAMP phosphodiesterases (PDEs). Several hypotheses have been proposed regarding the mechanisms by which the cAMP level is influenced in PKD. At first, reduced calcium directly inhibits PDE1, indirectly suppresses PDE3, and activates membrane bound AC6 (Gattone et al. 2003; Wang et al. 2010). Next, a defect in PC2-mediated calcium entry in the ciliary protein complex occurs, leading to the inhibition of AC5/6 and the activation of PDE4C (Choi et al. 2011). Also, depletion of the endoplasmic reticulum (ER) calcium store triggers the accumulation of stromal interaction molecule 1 to the plasma membrane and activates AC6 (Spirli et al. 2012). Likewise, several factors can contribute to an increase in the intracellular level of cAMP—disrupted PC1 binds to heterotrimeric G proteins (Parnell et al. 1998), vasopressin V2 receptor is upregulated, and circulating vasopressin, forskolin, ATP, or other adenylyl cyclase agonists are increased (Putnam et al. 2007; Hovater et al. 2008). Certain ACs and PDEs are recognized as being important in PKD progression because of their influence on compartmentalized pools of cAMP (Torres and Harris 2014).

One of the most evident characteristics of ADPKD pathogenesis is elevated cellular growth and division. Polycystin proteins inhibit cell growth through interactions with several pathways including the mTOR (Shillingford et al. 2006) and Janus kinase (JAK)-signal transducers and activators of transcription (STAT) (Bhunia et al. 2002) pathways. PC1 inhibits mTOR activity by stabilizing the tuberous sclerosis 1-tuberous sclerosis 2 (TSC1-TSC2) complex, which is known as a negative regulator of the mTOR complex (Huang and Manning 2008; Distefano et al. 2009; Dere et al. 2010). PC1 stabilizes the TSC1-TSC2 complex through two distinct mechanisms. PC1 suppresses the ERK-dependent phosphorylation of TSC2 at S664 (Distefano et al. 2009) and Akt-dependent phosphorylation of TSC2 at S939 by binding to TSC2 at the plasma membrane (Dere et al. 2010). The influence of the PC1-TSC2 interaction allows the TSC1-TSC2 complex to inhibit the mTOR signaling pathway. PC1 also functions as a positive regulator of p21 (cyclin-dependent kinase inhibitor) by binding and activating members of the JAK-STAT pathway (Bhunia et al. 2002). Following the PC2-JAK2 interaction and the formation of the intact C terminus of PC1, PC1 can activate STAT1 and STAT3 and it allow to increase p21 and decrease cell growth (Bhunia et al. 2002). PC2 also reduces cell proliferation by binding both eukaryotic translation elongation initiation factor 2a (eIF2a) and pancreatic ER-resident eIF2a kinase (Liang et al. 2008).

The Wnt signaling pathways regulate cell growth, differentiation, and planar cell polarity and are classified into the canonical (β-catenin dependent) and noncanonical (β-catenin independent) pathways. Both PC1 and PC2 affect the canonical Wnt pathway. In the case of PC1, the cleaved PC1 CTT directly or indirectly binds to β-catenin, translocates to the nucleus, and promotes T cell factor (TCF)-dependent transcription (Lal et al. 2008). PC2 also modulates the expression levels of some Wnt pathway components (Kim et al. 2009). In the noncanonical Wnt pathway, the function of PC1 is associated with the maintenance of planar cell polarity. Planar cell polarity is essential for oriented cell division and the establishment of kidney tubule structure, and defects in this process trigger the expansion of renal tubules and cyst formation (Fischer et al. 2006).

2.3 Genetic Mechanisms of Pathogenesis

Approximately 85 % of ADPKD patients have mutations in PKD1; thus, it is supposed that mutations in PKD1 cause a more severe disease than mutations in PKD2 do (Rossetti et al. 2007). Generally, patients with PKD1 mutations develop ADPKD symptoms at younger ages relative to patients with PKD2 mutations, but their disease phenotypes are influenced by mutations in both genes (Hateboer et al. 1999). A myriad of mutation types that can cause ADPKD have been revealed, and the position of each mutation determines the severity of the disease (Rossetti et al. 2002; Rossetti and Harris 2013). Recent studies revealed that the type of mutation is more important, because patients with truncating mutations showed more severe disease phenotypes than did those with non-truncating mutations (Pei et al. 2012; Cornec-Le Gall et al. 2013).

ADPKD is genetically dominant at the organismal level, but recessive at the cellular level (Chapin and Caplan 2010). Although a germ line mutation in PKD1 or PKD2 is necessary to induce cyst formation in ADPKD, cysts form in only part of the kidney tubules and hepatic bile duct. However, in adult tissues, both copies of the mutated polycystic gene undergo recessive loss of function, causing cyst formation to be accelerated in a subset of tubular epithelial cells. This paradox is explained by the occurrence of an additional somatic "second hit" mutation (Qian et al. 1996; Watnick et al. 1998; Pei et al. 1999). Although the somatic second hit mutation mechanism is generally applicable to human ADPKD, additional factors contribute to determining the extent of cyst formation, including non-cell autonomous effects on polycystins-expressing cells (Nishio et al. 2005), the timing of PKD1 in the developmental stages (Piontek et al. 2007), and hypomorphic mutations of PC1 compared to complete loss of function mutations (Rossetti et al. 2009; Hopp et al. 2012). Especially, reduced PC1 dosage is suggested to explain autosomal recessive PKD phenotypes, in which the degree of PC1 dysregulation is associated with the extent of tubule dilation and cyst formation (Hopp et al. 2012).

ADPKD is characterized by the formation of multiple fluid-filled kidney cysts. Thus, we need to focus on the mechanisms of cyst expansion. In the patient's kidney, cells are organized in a circle and these lumens must fill with fluid. Subsequently, cysts increase by cell proliferation, leading to the dilation of the renal tubule and renal failure (Qian et al. 1996; Brasier and Henske 1997). One model involves a loss of oriented cell division in the cells of mouse models with kidney-specific PKD1 or PKD2 mutation, which does not initiate cyst formation. This model suggests that a defect of planar cell polarity is an important factor in the expansion of many cysts; however, this factor was not essential for the initiation of cyst formation (Nishio et al. 2010). According to other studies, ion absorption and secretion in cyst-lining epithelial cells are significant for cyst formation. cAMP can stimulate Cl^- transport, resulting in the rapid and progressive dilation of tubules (Grantham 1996). In addition, one research group reported that tubule enlargement is prevented by inhibitors of Na^+–K^+–$2Cl^-$ cotransporters and CFTR (Montesano et al. 2009). Therefore, cAMP signaling plays a key role in renal cyst formation by promoting Cl^- driven fluid secretion. Furthermore, polycystin proteins also function as regulators of cAMP signaling by modulating the expression, localization, and activity of Cl^- channels (Chapin and Caplan 2010).

References

Anyatonwu GI, Ehrlich BE (2005) Organic cation permeation through the channel formed by polycystin-2. J Biol Chem 280(33):29488–29493. doi:10.1074/jbc.M504359200

Anyatonwu GI, Estrada M, Tian X, Somlo S, Ehrlich BE (2007) Regulation of ryanodine receptor-dependent calcium signaling by polycystin-2. Proc Natl Acad Sci U S A 104(15):6454–6459. doi:10.1073/pnas.0610324104

Babich V, Zeng WZ, Yeh BI, Ibraghimov-Beskrovnaya O, Cai Y, Somlo S, Huang CL (2004) The N-terminal extracellular domain is required for polycystin-1-dependent channel activity. J Biol Chem 279(24):25582–25589. doi:10.1074/jbc.M402829200

Banizs B, Komlosi P, Bevensee MO, Schwiebert EM, Bell PD, Yoder BK (2007) Altered pH(i) regulation and Na(+)/HCO3(−) transporter activity in choroid plexus of cilia-defective Tg737(orpk) mutant mouse. Am J Physiol Cell Physiol 292(4):C1409–C1416. doi:10.1152/ajpcell.00408.2006

Bhunia AK, Piontek K, Boletta A, Liu L, Qian F, Xu PN, Germino FJ, Germino GG (2002) PKD1 induces p21(waf1) and regulation of the cell cycle via direct activation of the JAK-STAT signaling pathway in a process requiring PKD2. Cell 109(2):157–168

Brasier JL, Henske EP (1997) Loss of the polycystic kidney disease (PKD1) region of chromosome 16p13 in renal cyst cells supports a loss-of-function model for cyst pathogenesis. J Clin Invest 99(2):194–199. doi:10.1172/JCI119147

Bycroft M, Bateman A, Clarke J, Hamill SJ, Sandford R, Thomas RL, Chothia C (1999) The structure of a PKD domain from polycystin-1: implications for polycystic kidney disease. EMBO J 18(2):297–305. doi:10.1093/emboj/18.2.297

Cai Y, Maeda Y, Cedzich A, Torres VE, Wu G, Hayashi T, Mochizuki T, Park JH, Witzgall R, Somlo S (1999) Identification and characterization of polycystin-2, the PKD2 gene product. J Biol Chem 274(40):28557–28565

Casuscelli J, Schmidt S, DeGray B, Petri ET, Celic A, Folta-Stogniew E, Ehrlich BE, Boggon TJ (2009) Analysis of the cytoplasmic interaction between polycystin-1 and polycystin-2. Am J Physiol Renal Physiol 297(5):F1310–F1315. doi:10.1152/ajprenal.00412.2009

Chapin HC, Caplan MJ (2010) The cell biology of polycystic kidney disease. J Cell Biol 191(4):701–710. doi:10.1083/jcb.201006173

Chauvet V, Qian F, Boute N, Cai Y, Phakdeekitacharoen B, Onuchic LF, Attie-Bitach T, Guicharnaud L, Devuyst O, Germino GG, Gubler MC (2002) Expression of PKD1 and PKD2 transcripts and proteins in human embryo and during normal kidney development. Am J Pathol 160(3):973–983. doi:10.1016/S0002-9440(10)64919-X

Choi YH, Suzuki A, Hajarnis S, Ma Z, Chapin HC, Caplan MJ, Pontoglio M, Somlo S, Igarashi P (2011) Polycystin-2 and phosphodiesterase 4C are components of a ciliary A-kinase anchoring protein complex that is disrupted in cystic kidney diseases. Proc Natl Acad Sci U S A 108(26):10679–10684. doi:10.1073/pnas.1016214108

Cornec-Le Gall E, Audrezet MP, Chen JM, Hourmant M, Morin MP, Perrichot R, Charasse C, Whebe B, Renaudineau E, Jousset P, Guillodo MP, Grall-Jezequel A, Saliou P, Ferec C, Le

Meur Y (2013) Type of PKD1 mutation influences renal outcome in ADPKD. J Am Soc Nephrol JASN 24(6):1006–1013. doi:10.1681/ASN.2012070650
Delmas P, Nomura H, Li X, Lakkis M, Luo Y, Segal Y, Fernandez-Fernandez JM, Harris P, Frischauf AM, Brown DA, Zhou J (2002) Constitutive activation of G-proteins by polycystin-1 is antagonized by polycystin-2. J Biol Chem 277(13):11276–11283. doi:10.1074/jbc.M110483200
Dere R, Wilson PD, Sandford RN, Walker CL (2010) Carboxy terminal tail of polycystin-1 regulates localization of TSC2 to repress mTOR. PLoS One 5(2):e9239. doi:10.1371/journal.pone.0009239
Distefano G, Boca M, Rowe I, Wodarczyk C, Ma L, Piontek KB, Germino GG, Pandolfi PP, Boletta A (2009) Polycystin-1 regulates extracellular signal-regulated kinase-dependent phosphorylation of tuberin to control cell size through mTOR and its downstream effectors S6K and 4EBP1. Mol Cell Biol 29(9):2359–2371. doi:10.1128/MCB.01259-08
Fischer E, Legue E, Doyen A, Nato F, Nicolas JF, Torres V, Yaniv M, Pontoglio M (2006) Defective planar cell polarity in polycystic kidney disease. Nat Genet 38(1):21–23. doi:10.1038/ng1701
Gallagher AR, Germino GG, Somlo S (2010) Molecular advances in autosomal dominant polycystic kidney disease. Adv Chronic Kidney Dis 17(2):118–130. doi:10.1053/j.ackd.2010.01.002
Gattone VH 2nd, Wang X, Harris PC, Torres VE (2003) Inhibition of renal cystic disease development and progression by a vasopressin V2 receptor antagonist. Nat Med 9(10):1323–1326. doi:10.1038/nm935
Gonzalez-Perrett S, Kim K, Ibarra C, Damiano AE, Zotta E, Batelli M, Harris PC, Reisin IL, Arnaout MA, Cantiello HF (2001) Polycystin-2, the protein mutated in autosomal dominant polycystic kidney disease (ADPKD), is a Ca2+-permeable nonselective cation channel. Proc Natl Acad Sci U S A 98(3):1182–1187
Grantham JJ (1996) The etiology, pathogenesis, and treatment of autosomal dominant polycystic kidney disease: recent advances. Am J Kidney Dis Off J Nat Kidney Found 28(6):788–803
Grimm DH, Cai Y, Chauvet V, Rajendran V, Zeltner R, Geng L, Avner ED, Sweeney W, Somlo S, Caplan MJ (2003) Polycystin-1 distribution is modulated by polycystin-2 expression in mammalian cells. J Biol Chem 278(38):36786–36793. doi:10.1074/jbc.M306536200
Hanaoka K, Qian F, Boletta A, Bhunia AK, Piontek K, Tsiokas L, Sukhatme VP, Guggino WB, Germino GG (2000) Co-assembly of polycystin-1 and -2 produces unique cation-permeable currents. Nature 408(6815):990–994. doi:10.1038/35050128
Harris PC, Ward CJ, Peral B, Hughes J (1995) Polycystic kidney disease. 1: identification and analysis of the primary defect. J Am Soc Nephrol JASN 6(4):1125–1133
Hateboer N, v Dijk MA, Bogdanova N, Coto E, Saggar-Malik AK, San Millan JL, Torra R, Breuning M, Ravine D (1999) Comparison of phenotypes of polycystic kidney disease types 1 and 2. European PKD1-PKD2 Study Group. Lancet 353(9147):103–107
Hidaka S, Konecke V, Osten L, Witzgall R (2004) PIGEA-14, a novel coiled-coil protein affecting the intracellular distribution of polycystin-2. J Biol Chem 279(33):35009–35016. doi:10.1074/jbc.M314206200
Hopp K, Ward CJ, Hommerding CJ, Nasr SH, Tuan HF, Gainullin VG, Rossetti S, Torres VE, Harris PC (2012) Functional polycystin-1 dosage governs autosomal dominant polycystic kidney disease severity. J Clin Invest 122(11):4257–4273. doi:10.1172/JCI64313
Hovater MB, Olteanu D, Welty EA, Schwiebert EM (2008) Purinergic signaling in the lumen of a normal nephron and in remodeled PKD encapsulated cysts. Purinergic Signal 4(2):109–124. doi:10.1007/s11302-008-9102-6
Huan Y, van Adelsberg J (1999) Polycystin-1, the PKD1 gene product, is in a complex containing E-cadherin and the catenins. J Clin Invest 104(10):1459–1468. doi:10.1172/JCI5111
Huang J, Manning BD (2008) The TSC1-TSC2 complex: a molecular switchboard controlling cell growth. Biochem J 412(2):179–190. doi:10.1042/BJ20080281
Ibraghimov-Beskrovnaya O, Dackowski WR, Foggensteiner L, Coleman N, Thiru S, Petry LR, Burn TC, Connors TD, Van Raay T, Bradley J, Qian F, Onuchic LF, Watnick TJ, Piontek K, Hakim RM, Landes GM, Germino GG, Sandford R, Klinger KW (1997) Polycystin: in vitro

synthesis, in vivo tissue expression, and subcellular localization identifies a large membrane-associated protein. Proc Natl Acad Sci U S A 94(12):6397–6402

Kim I, Ding T, Fu Y, Li C, Cui L, Li A, Lian P, Liang D, Wang DW, Guo C, Ma J, Zhao P, Coffey RJ, Zhan Q, Wu G (2009) Conditional mutation of Pkd2 causes cystogenesis and upregulates beta-catenin. J Am Soc Nephrol JASN 20(12):2556–2569. doi:10.1681/ASN.2009030271

Kip SN, Hunter LW, Ren Q, Harris PC, Somlo S, Torres VE, Sieck GC, Qian Q (2005) [Ca2+]i reduction increases cellular proliferation and apoptosis in vascular smooth muscle cells: relevance to the ADPKD phenotype. Circ Res 96(8):873–880. doi:10.1161/01.RES.0000163278.68142.8a

Koulen P, Cai Y, Geng L, Maeda Y, Nishimura S, Witzgall R, Ehrlich BE, Somlo S (2002) Polycystin-2 is an intracellular calcium release channel. Nat Cell Biol 4(3):191–197. doi:10.1038/ncb754

Lal M, Song X, Pluznick JL, Di Giovanni V, Merrick DM, Rosenblum ND, Chauvet V, Gottardi CJ, Pei Y, Caplan MJ (2008) Polycystin-1 C-terminal tail associates with beta-catenin and inhibits canonical Wnt signaling. Hum Mol Genet 17(20):3105–3117. doi:10.1093/hmg/ddn208

Li Y, Santoso NG, Yu S, Woodward OM, Qian F, Guggino WB (2009) Polycystin-1 interacts with inositol 1,4,5-trisphosphate receptor to modulate intracellular Ca2+ signaling with implications for polycystic kidney disease. J Biol Chem 284(52):36431–36441. doi:10.1074/jbc.M109.068916

Liang G, Yang J, Wang Z, Li Q, Tang Y, Chen XZ (2008) Polycystin-2 down-regulates cell proliferation via promoting PERK-dependent phosphorylation of eIF2alpha. Hum Mol Genet 17(20):3254–3262. doi:10.1093/hmg/ddn221

Low SH, Vasanth S, Larson CH, Mukherjee S, Sharma N, Kinter MT, Kane ME, Obara T, Weimbs T (2006) Polycystin-1, STAT6, and P100 function in a pathway that transduces ciliary mechanosensation and is activated in polycystic kidney disease. Dev Cell 10(1):57–69. doi:10.1016/j.devcel.2005.12.005

Masyuk AI, Masyuk TV, Splinter PL, Huang BQ, Stroope AJ, LaRusso NF (2006) Cholangiocyte cilia detect changes in luminal fluid flow and transmit them into intracellular Ca2+ and cAMP signaling. Gastroenterology 131(3):911–920. doi:10.1053/j.gastro.2006.07.003

Masyuk TV, Masyuk AI, Torres VE, Harris PC, Larusso NF (2007) Octreotide inhibits hepatic cystogenesis in a rodent model of polycystic liver disease by reducing cholangiocyte adenosine 3′,5′-cyclic monophosphate. Gastroenterology 132(3):1104–1116. doi:10.1053/j.gastro.2006.12.039

Mochizuki T, Wu G, Hayashi T, Xenophontos SL, Veldhuisen B, Saris JJ, Reynolds DM, Cai Y, Gabow PA, Pierides A, Kimberling WJ, Breuning MH, Deltas CC, Peters DJ, Somlo S (1996) PKD2, a gene for polycystic kidney disease that encodes an integral membrane protein. Science 272(5266):1339–1342

Montesano R, Ghzili H, Carrozzino F, Rossier BC, Feraille E (2009) cAMP-dependent chloride secretion mediates tubule enlargement and cyst formation by cultured mammalian collecting duct cells. Am J Physiol Renal Physiol 296(2):F446–F457. doi:10.1152/ajprenal.90415.2008

Nauli SM, Alenghat FJ, Luo Y, Williams E, Vassilev P, Li X, Elia AE, Lu W, Brown EM, Quinn SJ, Ingber DE, Zhou J (2003) Polycystins 1 and 2 mediate mechanosensation in the primary cilium of kidney cells. Nat Genet 33(2):129–137. doi:10.1038/ng1076

Nishio S, Hatano M, Nagata M, Horie S, Koike T, Tokuhisa T, Mochizuki T (2005) Pkd1 regulates immortalized proliferation of renal tubular epithelial cells through p53 induction and JNK activation. J Clin Invest 115(4):910–918. doi:10.1172/JCI22850

Nishio S, Tian X, Gallagher AR, Yu Z, Patel V, Igarashi P, Somlo S (2010) Loss of oriented cell division does not initiate cyst formation. J Am Soc Nephrol JASN 21(2):295–302. doi:10.1681/ASN.2009060603

Parnell SC, Magenheimer BS, Maser RL, Rankin CA, Smine A, Okamoto T, Calvet JP (1998) The polycystic kidney disease-1 protein, polycystin-1, binds and activates heterotrimeric G-proteins in vitro. Biochem Biophys Res Commun 251(2):625–631. doi:10.1006/bbrc.1998.9514

Pei Y, Watnick T, He N, Wang K, Liang Y, Parfrey P, Germino G, St George-Hyslop P (1999) Somatic PKD2 mutations in individual kidney and liver cysts support a "two-hit" model of cystogenesis in type 2 autosomal dominant polycystic kidney disease. J Am Soc Nephrol JASN 10(7):1524–1529

Pei Y, Lan Z, Wang K, Garcia-Gonzalez M, He N, Dicks E, Parfrey P, Germino G, Watnick T (2012) A missense mutation in PKD1 attenuates the severity of renal disease. Kidney Int 81(4):412–417. doi:10.1038/ki.2011.370

Piontek K, Menezes LF, Garcia-Gonzalez MA, Huso DL, Germino GG (2007) A critical developmental switch defines the kinetics of kidney cyst formation after loss of Pkd1. Nat Med 13(12):1490–1495. doi:10.1038/nm1675

Putnam WC, Swenson SM, Reif GA, Wallace DP, Helmkamp GM Jr, Grantham JJ (2007) Identification of a forskolin-like molecule in human renal cysts. J Am Soc Nephrol JASN 18(3):934–943. doi:10.1681/ASN.2006111218

Qian F, Watnick TJ, Onuchic LF, Germino GG (1996) The molecular basis of focal cyst formation in human autosomal dominant polycystic kidney disease type I. Cell 87(6):979–987

Qian F, Germino FJ, Cai Y, Zhang X, Somlo S, Germino GG (1997) PKD1 interacts with PKD2 through a probable coiled-coil domain. Nat Genet 16(2):179–183. doi:10.1038/ng0697-179

Qian F, Boletta A, Bhunia AK, Xu H, Liu L, Ahrabi AK, Watnick TJ, Zhou F, Germino GG (2002) Cleavage of polycystin-1 requires the receptor for egg jelly domain and is disrupted by human autosomal-dominant polycystic kidney disease 1-associated mutations. Proc Natl Acad Sci U S A 99(26):16981–16986. doi:10.1073/pnas.252484899

Rossetti S, Harris PC (2013) The genetics of vascular complications in autosomal dominant polycystic kidney disease (ADPKD). Curr Hypertens Rev 9(1):37–43

Rossetti S, Burton S, Strmecki L, Pond GR, San Millan JL, Zerres K, Barratt TM, Ozen S, Torres VE, Bergstralh EJ, Winearls CG, Harris PC (2002) The position of the polycystic kidney disease 1 (PKD1) gene mutation correlates with the severity of renal disease. J Am Soc Nephrol JASN 13(5):1230–1237

Rossetti S, Consugar MB, Chapman AB, Torres VE, Guay-Woodford LM, Grantham JJ, Bennett WM, Meyers CM, Walker DL, Bae K, Zhang QJ, Thompson PA, Miller JP, Harris PC, Consortium C (2007) Comprehensive molecular diagnostics in autosomal dominant polycystic kidney disease. J Am Soc Nephrol JASN 18(7):2143–2160. doi:10.1681/ASN.2006121387

Rossetti S, Kubly VJ, Consugar MB, Hopp K, Roy S, Horsley SW, Chauveau D, Rees L, Barratt TM, Van't Hoff WG, Niaudet P, Torres VE, Harris PC (2009) Incompletely penetrant PKD1 alleles suggest a role for gene dosage in cyst initiation in polycystic kidney disease. Kidney Int 75(8):848–855. doi:10.1038/ki.2008.686

Shillingford JM, Murcia NS, Larson CH, Low SH, Hedgepeth R, Brown N, Flask CA, Novick AC, Goldfarb DA, Kramer-Zucker A, Walz G, Piontek KB, Germino GG, Weimbs T (2006) The mTOR pathway is regulated by polycystin-1, and its inhibition reverses renal cystogenesis in polycystic kidney disease. Proc Natl Acad Sci U S A 103(14):5466–5471. doi:10.1073/pnas.0509694103

Spirli C, Locatelli L, Fiorotto R, Morell CM, Fabris L, Pozzan T, Strazzabosco M (2012) Altered store operated calcium entry increases cyclic 3′,5′-adenosine monophosphate production and extracellular signal-regulated kinases 1 and 2 phosphorylation in polycystin-2-defective cholangiocytes. Hepatology 55(3):856–868. doi:10.1002/hep.24723

Streets AJ, Wagner BE, Harris PC, Ward CJ, Ong AC (2009) Homophilic and heterophilic polycystin 1 interactions regulate E-cadherin recruitment and junction assembly in MDCK cells. J Cell Sci 122(Pt 9):1410–1417. doi:10.1242/jcs.045021

Torres VE, Harris PC (2014) Strategies targeting cAMP signaling in the treatment of polycystic kidney disease. J Am Soc Nephrol JASN 25(1):18–32. doi:10.1681/ASN.2013040398

Tsiokas L, Kim E, Arnould T, Sukhatme VP, Walz G (1997) Homo- and heterodimeric interactions between the gene products of PKD1 and PKD2. Proc Natl Acad Sci U S A 94(13):6965–6970

Vassilev PM, Guo L, Chen XZ, Segal Y, Peng JB, Basora N, Babakhanlou H, Cruger G, Kanazirska M, Ye C, Brown EM, Hediger MA, Zhou J (2001) Polycystin-2 is a novel cation channel implicated in defective intracellular Ca(2+) homeostasis in polycystic kidney disease. Biochem Biophys Res Commun 282(1):341–350. doi:10.1006/bbrc.2001.4554

Wang X, Ward CJ, Harris PC, Torres VE (2010) Cyclic nucleotide signaling in polycystic kidney disease. Kidney Int 77(2):129–140. doi:10.1038/ki.2009.438

Ward CJ, Turley H, Ong AC, Comley M, Biddolph S, Chetty R, Ratcliffe PJ, Gattner K, Harris PC (1996) Polycystin, the polycystic kidney disease 1 protein, is expressed by epithelial cells in fetal, adult, and polycystic kidney. Proc Natl Acad Sci U S A 93(4):1524–1528

Watnick TJ, Torres VE, Gandolph MA, Qian F, Onuchic LF, Klinger KW, Landes G, Germino GG (1998) Somatic mutation in individual liver cysts supports a two-hit model of cystogenesis in autosomal dominant polycystic kidney disease. Mol Cell 2(2):247–251

Wei W, Hackmann K, Xu H, Germino G, Qian F (2007) Characterization of cis-autoproteolysis of polycystin-1, the product of human polycystic kidney disease 1 gene. J Biol Chem 282(30):21729–21737. doi:10.1074/jbc.M703218200

Xu C, Rossetti S, Jiang L, Harris PC, Brown-Glaberman U, Wandinger-Ness A, Bacallao R, Alper SL (2007) Human ADPKD primary cyst epithelial cells with a novel, single codon deletion in the PKD1 gene exhibit defective ciliary polycystin localization and loss of flow-induced Ca2+ signaling. Am J Physiol Renal Physiol 292(3):F930–F945. doi:10.1152/ajprenal.00285.2006

Yoder BK, Hou X, Guay-Woodford LM (2002) The polycystic kidney disease proteins, polycystin-1, polycystin-2, polaris, and cystin, are co-localized in renal cilia. J Am Soc Nephrol JASN 13(10):2508–2516

Yu S, Hackmann K, Gao J, He X, Piontek K, Garcia-Gonzalez MA, Menezes LF, Xu H, Germino GG, Zuo J, Qian F (2007) Essential role of cleavage of Polycystin-1 at G protein-coupled receptor proteolytic site for kidney tubular structure. Proc Natl Acad Sci U S A 104(47):18688–18693. doi:10.1073/pnas.0708217104

Yu Y, Ulbrich MH, Li MH, Buraei Z, Chen XZ, Ong AC, Tong L, Isacoff EY, Yang J (2009) Structural and molecular basis of the assembly of the TRPP2/PKD1 complex. Proc Natl Acad Sci U S A 106(28):11558–11563. doi:10.1073/pnas.0903684106

Part II
Cystogenesis Mechanisms

Chapter 3
Cell Proliferation and Apoptosis in ADPKD

Eun Ji Lee

Abstract Increased tubular epithelial cell proliferation with fluid secretion is a key hallmark of autosomal dominant polycystic kidney disease (ADPKD). With disruption of either *PKD1* or *PKD2*, the main causative genes of ADPKD, intracellular calcium homeostasis and cAMP accumulation are disrupted, which in turn leads to altered signaling in the pathways that regulate cell proliferation. These dysregulations finally stimulate the development of fluid-filled cysts originating from abnormally proliferating renal tubular cells. In addition, dysregulated apoptosis is observed in dilated cystic tubules. An imbalance between cell proliferation and apoptosis seems to contribute to cyst growth and renal tissue remodeling in ADPKD. In this section, the mechanisms through which cell proliferation and apoptosis are involved in disease progression, and further, how those signaling pathways impinge on each other in ADPKD will be discussed.

Keywords Apoptosis • Proliferation • Autosomal dominant polycystic kidney disease • ADPKD

3.1 Cell Proliferation in ADPKD

Cell proliferation is an important intracellular process in which nearly all of the billions of cells in our body undergo in a strictly regulated manner. It allows cell populations to increase through cell growth and division. Mechanisms that regulate cell proliferation include cell cycle controls and series of protein kinase cascades stimulated by various growth factors. The cell cycle mainly controls cell division and it can be divided into four stages: the G1, S, G2, and M phases. DNA replication and the duplication of identical sets of chromosomes occurs during S phase. Cell division finally occurs in M phase and involves DNA packing and chromosome segregation. Gap phases between the S and M phases involve preparation for the following

E.J. Lee (✉)
Molecular Medicine Laboratory, Department of Life systems, Sookmyung Women's University, Cheongpa-ro 47-gil 100, Yongsan-gu, Seoul 04310, South Korea
e-mail: eunji8902@sm.ac.kr

J.H. Park, C. Ahn (eds.), *Cystogenesis*, Advances in Experimental Medicine and Biology 933, DOI 10.1007/978-981-10-2041-4_3

stages and the determination of whether to proceed to the next stage. The cell cycle is mainly mediated by cyclin-dependent kinases (CDKs), which are positive regulators of the cell cycle, and their inhibitors. Those proteins control cell cycle through switching the activity of specific proteins on and off by phosphorylation at appropriate time points in the cell cycle (Mitchison 2003). Not only cell cycle control, but also several signaling pathways mediated by series of protein kinases are well-characterized mechanisms regulating cell proliferation. One of the best understood signaling pathways of the mitogen-activated protein kinases (MAPKs) is the extracellular signal-regulated kinase (ERK) pathway. Binding of growth factors or cytokines to their cognate receptors sequentially activates MAPKs in a multi-step process. Signaling occurs through a cascade of protein kinases including B-Raf, mitogene-activated protein kinase kinase (MEK), and ERK. Ultimately, phosphorylated ERK translocates to the nucleus and activates transcription factors, thereby altering gene expression to stimulate cell proliferation (Zhang and Liu 2002). Another mechanism by which protein kinases influence the control of cell growth is via the mammalian target of rapamycin (mTOR) signaling pathway, which can be stimulated by extracellular factors including growth factors. When it is activated, mTOR phosphorylates the serine/threonine protein kinase Akt and finally leads to enhanced cell proliferation (Dobashi et al. 2011).

In normal tissues that are fully differentiated, cell proliferation rarely occurs; in such tissues the cells each have their own specialized function and no longer actively divide. Therefore, cell proliferation is tightly controlled and it is highly restricted to cells that mediate replenishment of the tissues under specific conditions, such as following physical injuries. Uncontrolled cell proliferation is abnormal in terminally differentiated tissues and is commonly observed in human diseases including a number of cancers (Evan and Vousden 2001), as well as being a characteristic hallmark of ADPKD.

3.1.1 Aberrant Calcium Signaling and Cell Proliferation in ADPKD

Polycystin-1 (PC1) and polycystin-2 (PC2), which are encoded by *PKD1* and *PKD2*, respectively, are localized in the primary cilia. PC1 and PC2 form a polcystin complex and function as a mechanosensor, translating mechanistic stimuli into calcium signaling. Specifically, PC2 acts as a nonselective cation channel and regulates the intracellular calcium level by transporting calcium into the cell (Gonzalez-Perrett et al. 2001). PC1, which is a large integral membrane receptor, interacts with PC2 and regulates its activity. Polycystin complex not only works in the cilia, but it can also regulate the intracellular calcium level by transporting calcium across the endoplasmic reticulum (ER) membrane. PC2, which is located on the ER membrane, stimulates inositol trisphosphate receptor (IP3R), leading to calcium release from the ER into the cytoplasm (Li et al. 2005). PC1, on the other hand, reduces the intracellular calcium level via inhibiting the interaction of PC2 with IP3R (Santoso et al. 2011). Above all, the disruption of polycystin complex proteins impairs intracellular calcium homeostasis, which leads to the alteration of various signaling pathways. The consequences of this include abnormally hyper-activated cell proliferation and fluid secretion, which leads to the expansion of renal cysts (Harris and Torres 2009).

Reduced intracellular calcium caused by defects in polycystin complex proteins primarily leads to cyclic adenosine monophosphate (cAMP) accumulation (Yamaguchi et al. 2004; Chebib et al. 2015). Previous studies revealed that calcium restriction is involved in both the synthesis and hydrolysis of cAMP. A low level of calcium stimulates the enzyme that catalyzes cAMP formation, adenylyl cyclase 6, which is essentially inhibited by calcium. Besides, the hydrolysis of cAMP is reduced via the inhibition of calcium-calmodulin dependent phosphodiesterases as a result of a low level of intracellular calcium (Wang et al. 2010). cAMP can also be modulated by circulating vasopressin. A high urinary concentration of vasopressin, which is commonly observed in ADPKD, increases the intracellular level of cAMP, mediated by vasopressin V2 receptor activation (Wang et al. 2008).

The accumulation of intracellular cAMP stimulates the cAMP-dependent B-Raf/MEK/ERK signaling pathway, which is one of the key regulators of cell proliferation (Yamaguchi et al. 2004). Hyper-activated ERK also affects the up-regulation of the mTOR signaling pathway, another representative pathway regulating cell proliferation, via the inhibition of tuberous sclerosis complex (TSC1/2) (Yamaguchi et al. 2003). Moreover, the activity of cAMP response element binding protein (CREB) is increased by cAMP and causes the over-expression of the epidermal growth factor (EGF)-like peptide amphiregulin, thereby leading to increased cell proliferation in an EGF receptor-dependent manner. CREB is also sequentially involved in the Raf/MEK/ERK pathway.

On the other hand, calcium influx can also be regulated by other heteromultimeric channels, including PC2/TRPC1 and PC2/TRPV4. Those channels are also localized on the cilia and transduce calcium into cells in response to fluid-flow (Bai et al. 2008). However, defects in either TRPC1 or TRPV4 do not lead to ADPKD, suggesting that an abnormal level of intracellular calcium is not the only relevant factor for triggering the development of renal cysts. Instead, the onset of ADPKD must be closely linked to dysfunctions in either PC1 or PC2, which are the main causative proteins of the disease (Ma et al. 2013).

3.2 Regulation of Cell Proliferation by Polycystins in Renal Primary Cilia

Cell proliferation is able to be regulated by polycystins directly as well as indirectly via a low intracellular calcium level.

First, PC1 inhibits the mTOR activity of normal renal epithelial cells via modulating the stability of TSC1/2, the negative regulator of mTOR (Distefano et al. 2009). The signaling pathway mediated by mTOR plays a central role in cell proliferation in response to growth factors including EGF and insulin-like growth factor 1. The activation of phosphatidylinositide 3-kinases followed by the stimulation of growth factor receptors leads to the sequential phosphorylation of protein kinase B (also known as Akt) and mTOR complex. Activated mTOR complex positively regulates protein synthesis (which finally allows enhanced cell proliferation) by the up- or down-regulation of ribosomal protein S6 kinase beta-1 or eukaryotic translation initiation factor 4E-binding protein 1, respectively. GTP-bound active Rheb, which is one of the subtypes of the Ras superfamily, is required in this process (Ruvinsky and Meyuhas 2006; Long et al. 2005; Laplante and Sabatini 2009). TSC1/2 complex inhibits the activation of Rheb. Therefore, the up-regulation of TSC1/2 by PC1 blocks mTOR activation and finally leads to the inhibition of cell proliferation in normal renal epithelia.

Another mechanism by which PC1 negatively regulates cell proliferation in normal conditions is through control of the cell cycle process. PC1 directly activates the Janus kinase and signal transducer and activator of transcription signaling pathway, leading to an increased expression of p21Waf1. p21Waf1 is a CDK inhibitor that causes cell cycle arrest in G0/G1 phase when it is expressed at a high level. A functionally intact interaction of PC1 with PC2 is necessary in this process, and mutations in PC1 or PC2 that affect the function of these proteins or inhibit their binding to each other usually results in dysregulated cell growth (Bhunia et al. 2002).

PC2 negatively regulates cell proliferation not only via its interaction with PC1, but also independently of PC1 via modulating the pancreatic ER-resident eukaryotic initiation factor 2 (eIF2) alpha kinase (PERK)-eIF2 phosphorylation signaling pathway. eIF2 is required to initiate translation by mediating the binding of tRNA to the ribosome, and through this activity it ultimately regulates cell growth control. The activity of eIF2 can be regulated by several factors including PERK, which is essentially induced by ER stress (Joshi et al. 2013). PC2 functions as a PERK-activating factor. It is physically localized within the PERK-eIF2 protein complex and it enhances eIF2 phosphorylation by PERK, which finally leads to restricted cell proliferation (Liang et al. 2008).

Altogether, in ADPKD, which is mainly caused by a defect in either PC1 or PC2, an abnormal increase of cell proliferation caused by several mechanisms is a hallmark of the disease. The inhibition of calcium influx that results from dysfunctions in the PC1-PC2 complex increases intracellular cAMP accumulation by both enhanced cAMP synthesis and reduced cAMP hydrolysis. This deficit in calcium transport directly or indirectly stimulates the Raf/MEK/ERK signaling pathways and finally leads to abnormally enhanced cell proliferation. Moreover, PC1 and PC2 essentially have roles in the blockade of cell proliferation via the regulation of key factors for cell proliferation. Thereby, the malfunction of either PC1 or PC2 results in increased cell proliferation. Indeed, the onset of ADPKD must be tightly related to hyper-activated cell proliferation.

3.3 Apoptosis in ADPKD

Apoptosis is a process of programmed cell death that eliminates damaged cells. Whereas necrosis is not uniformly regulated, apoptosis is highly controlled by its regulators and undergoes defined sequential changes in morphology. These changes include a reduced cell volume, membrane blebbing, and loss of cellular membrane asymmetry, as well as nuclear fragmentation, chromatin condensation, and DNA fragmentation. During this process, phosphatidylserine is exposed at the surface of the cell and promotes the phagocytosis of the apoptotic cell by immune cells (Elmore 2007). The major regulators of apoptosis are the caspases and inhibitor of apoptosis proteins (IAPs). Caspases, which are cysteine proteases, exist in the cytoplasm as inactive forms in normal cells. Upon the activation of apoptosis, caspases are transformed into their functional forms via two serial cleavage steps and initiate apoptosis (Wolf and Green 1999). IAPs are the natural inhibitors of caspases and they antagonize apoptosis through inhibiting the action of caspases as well as regulating cell division. There are two representative activation mechanisms of apoptosis, which are known as the extrinsic and intrinsic pathways. The intrinsic pathway, also known as the mitochondrial pathway, is mainly induced by increased cytochrome c release from mitochondria. Cytochrome c interacts with apoptosis protease-activating factor-1 to serially recruit caspases including caspase-9, caspase-3, and caspase-7; thereby, apoptosis signaling is initiated. In the extrinsic pathway, extracellular factors, rather than cytosolic ones, are involved in the initial step of apoptosis activation. After the binding of ligands to death receptors (such as the FAS-FAS receptor), caspases are recruited and cleaved into their active forms followed by their recruitment by adaptor proteins. The caspase that mainly mediates extrinsic apoptosis is caspase-8, and it sequentially activates other caspases, resulting in an activation cascade (Ashkenazi and Dixit 1998; Portt et al. 2011). Apoptosis regulated both by intrinsic and extrinsic mechanisms has a critical role in normal kidney development and its dysregulation has been reported to be related to various types of renal diseases including cancer and ADPKD (Goilav 2011).

3.3.1 Altered Regulation of Apoptosis During ADPKD Pathogenesis

The dysregulation of apoptosis is commonly observed in rodent models of ADPKD. First, apoptosis has been reported to be up-regulated in the Han:SPRD rat model, which is derived from a spontaneous mutation in the Sprague–Dawley strain and has polycystic kidney disease (PKD) phenotypes. Homozygous animals that aggressively develop renal cysts have shown increased activities of apoptosis-inducing factors including caspases (specifically caspase-3, caspase-7, caspase-8, and caspase-2) as well as enhanced cytochrome c release from mitochondria into the cytoplasm (Tao et al. 2005). Conversely, PKD phenotypes have appeared in rodent models that have either an increased or decreased genetic expression of apoptosis-related genes. Transgenic mice targeting c-Myc, a proto-oncogene that is involved in cell proliferation and apoptosis, have been found to develop polycystic kidneys. These mice finally develop end-stage renal disease and die of renal failure (Lanoix et al. 1996). Besides, an in vivo model that has a genetic knock-out of pro-apoptotic Bcl-2 also showed PKD phenotypes and hyper-activation of cell proliferation and apoptosis (Veis et al. 1993; Sorenson et al. 1996). Interestingly, the level of those PKD-causing apoptosis related genes turned out to be increased in ADPKD, while others including Bax and P53 were unchanged (Lanoix et al. 1996). Other observations of human ADPKD have shown that increased apoptosis is specifically detected only in the renal epithelial tubules of the non-cystic region in patient tissues. Indeed, apoptosis may be involved in the loss of normal nephrons, leading to the destruction of renal architecture, rather than having a major role in cyst development (Woo 1995).

Assertions for the role of apoptosis in ADPKD are controversial (Zhou and Kukes 1998). Some consider dysregulated apoptosis to be one of the main causes of ADPKD, whereas others have proposed that apoptosis may delay disease progression. Clearly, the hyper-activation of cell proliferation, which is caused by increased intracellular cAMP followed by the inhibition of calcium influx due to the dysfunction of the PC1-PC2 complex, is the core factor for ADPKD onset. However, it is evident that abnormal apoptosis has also been reported in several studies; thereby, the failure to maintain a proper balance between proliferation and apoptosis seems to be important in progression of ADPKD (Ecder et al. 2002).

3.4 Therapeutic Approaches Targeting Cell Proliferation or Apoptosis

No specific therapies or medications are available to prevent or delay ADPKD progression yet. However, there have been several trials of therapies targeting either cell proliferation or apoptosis, which are the main causative mechanisms inducing cyst growth and fluid secretion (LaRiviere et al. 2015; Riella et al. 2014). V2 receptor antagonists are one drug that has been investigated as a therapy for ADPKD. V2 receptor is stimulated by vasopressin, and intracellular signaling pathways mediated by vasopressin-V2 receptor signaling are highly involved in cAMP generation. In pre-clinical attempts, ADPKD rodent models treated by antagonists targeting V2 receptor showed alleviated disease progression (Gattone et al. 2003; Torres et al. 2004; Wang et al. 2005). Another trial drug targeting the cAMP signaling pathway was somatostatin analogs. Essentially, somatostatin binds with several G-protein coupled receptors (GPCRs) and thereby maintains low cAMP levels via the regulation of ACs activity. It also down-regulates cell proliferation and the secretion of hormones or growth factors. Therefore, somatostatin analogs lowered intracellular cAMP, which finally leads to attenuated cystogenesis followed by the down-regulation of cell proliferation, in preclinical studies using rodent models with renal cysts

(Masyuk et al. 2007; Masyuk et al. 2013). Besides, inhibitors targeting the mTOR signaling pathway, which is a major mechanism regulating cell proliferation, have been shown to effectively attenuate PKD phenotypes including renal enlargement and development of the cysts (Wahl et al. 2006) through in vivo studies. Pre-clinical trials of drugs targeting apoptosis have been attempted relatively less frequently than those of drugs targeting cell proliferation. The down-regulation of apoptosis followed by caspase inhibition has revealed alleviated PKD progression in a rodent model with renal cysts. Additionally, a drug that directly targets the potential apoptosis-inducing factor CDK5 was indeed shown to inhibit cystic phenotypes in a rodent model of PKD (Bukanov et al. 2006; Ibraghimov-Beskrovnaya 2007).

3.5 Summary

In summary, the hyper-activation of cell proliferation is a key phenotype that initiates and accelerates the progression of ADPKD. It could be explained by two major mechanisms. The first mechanism is cAMP accumulation followed by lowered calcium influx, which stimulates the activities of cell proliferation-related proteins including B-Raf/MEK/ERK and mTOR. The second mechanism is the loss of the anti-proliferative functions of PC1 and PC2 through defects in either protein caused by gene mutations. The role of excessive cell proliferation in ADPKD has been clearly characterized and pre-clinical trials of drugs targeting cell proliferation have reported the effective attenuation of disease progression. On the other hand, the role of apoptosis in ADPKD pathogenesis is not yet fully understood. Apoptosis regulators including caspases have recently been reported to show increased activity in ADPKD rodent models, and indirectly targeting apoptosis in pre-clinical studies has been found to delay disease onset. Besides, renal failure followed by cysts development has been commonly observed in mice with defective apoptosis. Taken together, those findings suggest that apoptosis seems to be involved in the pathogenesis of ADPKD, associated either with cysts enlargement or the remodeling of kidney structure. Altogether, the failure to maintain an appropriate balance between cell proliferation and apoptosis appears to be an important driver of ADPKD progression. Therefore, specifically targeting proliferation and/or apoptosis could be an effective therapeutic strategy.

References

Ashkenazi A, Dixit VM (1998) Death receptors: signaling and modulation. Science 281(5381):1305–1308

Bai CX, Giamarchi A, Rodat-Despoix L, Padilla F, Downs T, Tsiokas L, Delmas P (2008) Formation of a new receptor-operated channel by heteromeric assembly of TRPP2 and TRPC1 subunits. EMBO Rep 9(5):472–479. doi:10.1038/embor.2008.29

Bhunia AK, Piontek K, Boletta A, Liu L, Qian F, Xu PN, Germino FJ, Germino GG (2002) PKD1 induces p21(waf1) and regulation of the cell cycle via direct activation of the JAK-STAT signaling pathway in a process requiring PKD2. Cell 109(2):157–168

Bukanov NO, Smith LA, Klinger KW, Ledbetter SR, Ibraghimov-Beskrovnaya O (2006) Long-lasting arrest of murine polycystic kidney disease with CDK inhibitor roscovitine. Nature 444(7121):949–952. doi:10.1038/nature05348

Chebib FT, Sussman CR, Wang X, Harris PC, Torres VE (2015) Vasopressin and disruption of calcium signalling in polycystic kidney disease. Nat Rev Nephrol 11(8):451–464. doi:10.1038/nrneph.2015.39

Distefano G, Boca M, Rowe I, Wodarczyk C, Ma L, Piontek KB, Germino GG, Pandolfi PP, Boletta A (2009) Polycystin-1 regulates extracellular signal-regulated kinase-dependent phosphorylation of tuberin to control cell size through mTOR and its downstream effectors S6K and 4EBP1. Mol Cell Biol 29(9):2359–2371. doi:10.1128/MCB.01259-08

Dobashi Y, Watanabe Y, Miwa C, Suzuki S, Koyama S (2011) Mammalian target of rapamycin: a central node of complex signaling cascades. Int J Clin Exp Pathol 4(5):476–495

Ecder T, Melnikov VY, Stanley M, Korular D, Lucia MS, Schrier RW, Edelstein CL (2002) Caspases, Bcl-2 proteins and apoptosis in autosomal-dominant polycystic kidney disease. Kidney Int 61(4):1220–1230. doi:10.1046/j.1523-1755.2002.00250.x

Elmore S (2007) Apoptosis: a review of programmed cell death. Toxicol Pathol 35(4):495–516. doi:10.1080/01926230701320337

Evan GI, Vousden KH (2001) Proliferation, cell cycle and apoptosis in cancer. Nature 411(6835):342–348. doi:10.1038/35077213

Gattone VH 2nd, Wang X, Harris PC, Torres VE (2003) Inhibition of renal cystic disease development and progression by a vasopressin V2 receptor antagonist. Nat Med 9(10):1323–1326. doi:10.1038/nm935

Goilav B (2011) Apoptosis in polycystic kidney disease. Biochim Biophys Acta 1812(10):1272–1280. doi:10.1016/j.bbadis.2011.01.006

Gonzalez-Perrett S, Kim K, Ibarra C, Damiano AE, Zotta E, Batelli M, Harris PC, Reisin IL, Arnaout MA, Cantiello HF (2001) Polycystin-2, the protein mutated in autosomal dominant polycystic kidney disease (ADPKD), is a Ca2+-permeable nonselective cation channel. Proc Natl Acad Sci U S A 98(3):1182–1187

Harris PC, Torres VE (2009) Polycystic kidney disease. Annu Rev Med 60:321–337. doi:10.1146/annurev.med.60.101707.125712

Ibraghimov-Beskrovnaya O (2007) Targeting dysregulated cell cycle and apoptosis for polycystic kidney disease therapy. Cell Cycle 6(7):776–779

Joshi M, Kulkarni A, Pal JK (2013) Small molecule modulators of eukaryotic initiation factor 2alpha kinases, the key regulators of protein synthesis. Biochimie 95(11):1980–1990. doi:10.1016/j.biochi.2013.07.030

Lanoix J, D'Agati V, Szabolcs M, Trudel M (1996) Dysregulation of cellular proliferation and apoptosis mediates human autosomal dominant polycystic kidney disease (ADPKD). Oncogene 13(6):1153–1160

Laplante M, Sabatini DM (2009) mTOR signaling at a glance. J Cell Sci 122(Pt 20):3589–3594. doi:10.1242/jcs.051011

LaRiviere WB, Irazabal MV, Torres VE (2015) Novel therapeutic approaches to autosomal dominant polycystic kidney disease. Transl Res J Lab Clin Med 165(4):488–498. doi:10.1016/j.trsl.2014.11.003

Li Y, Wright JM, Qian F, Germino GG, Guggino WB (2005) Polycystin 2 interacts with type I inositol 1,4,5-trisphosphate receptor to modulate intracellular Ca2+ signaling. J Biol Chem 280(50):41298–41306. doi:10.1074/jbc.M510082200

Liang G, Yang J, Wang Z, Li Q, Tang Y, Chen XZ (2008) Polycystin-2 down-regulates cell proliferation via promoting PERK-dependent phosphorylation of eIF2alpha. Hum Mol Genet 17(20):3254–3262. doi:10.1093/hmg/ddn221

Long X, Lin Y, Ortiz-Vega S, Yonezawa K, Avruch J (2005) Rheb binds and regulates the mTOR kinase. Curr Biol CB 15(8):702–713. doi:10.1016/j.cub.2005.02.053
Ma M, Tian X, Igarashi P, Pazour GJ, Somlo S (2013) Loss of cilia suppresses cyst growth in genetic models of autosomal dominant polycystic kidney disease. Nat Genet 45(9):1004–1012. doi:10.1038/ng.2715
Masyuk TV, Masyuk AI, Torres VE, Harris PC, Larusso NF (2007) Octreotide inhibits hepatic cystogenesis in a rodent model of polycystic liver disease by reducing cholangiocyte adenosine 3′,5′-cyclic monophosphate. Gastroenterology 132(3):1104–1116. doi:10.1053/j.gastro.2006.12.039
Masyuk TV, Radtke BN, Stroope AJ, Banales JM, Gradilone SA, Huang B, Masyuk AI, Hogan MC, Torres VE, Larusso NF (2013) Pasireotide is more effective than octreotide in reducing hepatorenal cystogenesis in rodents with polycystic kidney and liver diseases. Hepatology 58(1):409–421. doi:10.1002/hep.26140
Mitchison JM (2003) Growth during the cell cycle. Int Rev Cytol 226:165–258
Portt L, Norman G, Clapp C, Greenwood M, Greenwood MT (2011) Anti-apoptosis and cell survival: a review. Biochim Biophys Acta 1813(1):238–259. doi:10.1016/j.bbamcr.2010.10.010
Riella C, Czarnecki PG, Steinman TI (2014) Therapeutic advances in the treatment of polycystic kidney disease. Nephron Clin Pract 128(3-4):297–302. doi:10.1159/000368244
Ruvinsky I, Meyuhas O (2006) Ribosomal protein S6 phosphorylation: from protein synthesis to cell size. Trends Biochem Sci 31(6):342–348. doi:10.1016/j.tibs.2006.04.003
Santoso NG, Cebotaru L, Guggino WB (2011) Polycystin-1, 2, and STIM1 interact with IP(3)R to modulate ER Ca release through the PI3K/Akt pathway. Cell Physiol Biochem Int J Exp Cell Physiol Biochem Pharmacol 27(6):715–726. doi:10.1159/000330080
Sorenson CM, Padanilam BJ, Hammerman MR (1996) Abnormal postpartum renal development and cystogenesis in the bcl-2 (-/-) mouse. Am J Phys 271(1 Pt 2):F184–F193
Tao Y, Kim J, Stanley M, He Z, Faubel S, Schrier RW, Edelstein CL (2005) Pathways of caspase-mediated apoptosis in autosomal-dominant polycystic kidney disease (ADPKD). Kidney Int 67(3):909–919. doi:10.1111/j.1523-1755.2005.00155.x
Torres VE, Wang X, Qian Q, Somlo S, Harris PC, Gattone VH 2nd (2004) Effective treatment of an orthologous model of autosomal dominant polycystic kidney disease. Nat Med 10(4):363–364. doi:10.1038/nm1004
Veis DJ, Sorenson CM, Shutter JR, Korsmeyer SJ (1993) Bcl-2-deficient mice demonstrate fulminant lymphoid apoptosis, polycystic kidneys, and hypopigmented hair. Cell 75(2):229–240
Wahl PR, Serra AL, Le Hir M, Molle KD, Hall MN, Wuthrich RP (2006) Inhibition of mTOR with sirolimus slows disease progression in Han: SPRD rats with autosomal dominant polycystic kidney disease (ADPKD). Nephrol Dial Transplant Off Publ Eur Dial Transplant Assoc Eur Renal Assoc 21(3):598–604. doi:10.1093/ndt/gfi181
Wang X, Gattone V 2nd, Harris PC, Torres VE (2005) Effectiveness of vasopressin V2 receptor antagonists OPC-31260 and OPC-41061 on polycystic kidney disease development in the PCK rat. J Am Soc Nephrol JASN 16(4):846–851. doi:10.1681/ASN.2004121090
Wang X, Wu Y, Ward CJ, Harris PC, Torres VE (2008) Vasopressin directly regulates cyst growth in polycystic kidney disease. J Am Soc Nephrol JASN 19(1):102–108. doi:10.1681/ASN.2007060688
Wang X, Ward CJ, Harris PC, Torres VE (2010) Cyclic nucleotide signaling in polycystic kidney disease. Kidney Int 77(2):129–140. doi:10.1038/ki.2009.438
Wolf BB, Green DR (1999) Suicidal tendencies: apoptotic cell death by caspase family proteinases. J Biol Chem 274(29):20049–20052
Woo D (1995) Apoptosis and loss of renal tissue in polycystic kidney diseases. N Engl J Med 333(1):18–25. doi:10.1056/NEJM199507063330104
Yamaguchi T, Nagao S, Wallace DP, Belibi FA, Cowley BD, Pelling JC, Grantham JJ (2003) Cyclic AMP activates B-Raf and ERK in cyst epithelial cells from autosomal-dominant polycystic kidneys. Kidney Int 63(6):1983–1994. doi:10.1046/j.1523-1755.2003.00023.x

Yamaguchi T, Wallace DP, Magenheimer BS, Hempson SJ, Grantham JJ, Calvet JP (2004) Calcium restriction allows cAMP activation of the B-Raf/ERK pathway, switching cells to a cAMP-dependent growth-stimulated phenotype. J Biol Chem 279(39):40419–40430. doi:10.1074/jbc.M405079200

Zhang W, Liu HT (2002) MAPK signal pathways in the regulation of cell proliferation in mammalian cells. Cell Res 12(1):9–18. doi:10.1038/sj.cr.7290105

Zhou XJ, Kukes G (1998) Pathogenesis of autosomal dominant polycystic kidney disease: role of apoptosis. Diagn Mol Pathol Am J Sur Pathol B 7(2):65–68

Chapter 4
Inflammation and Fibrosis in ADPKD

Hyowon Mun and Jong Hoon Park

Abstract Diverse signaling pathways have been reported to be associated with polycystic kidney disease (PKD). Cell proliferation is widely known to be an important pathway related to this disease. However, studies on the interactions of inflammation and fibrosis with polycystic kidney disease have been limited. Inflammation is one of the protective systems involved in the response to foreign molecules. In PKD, it was reported that the activity of signaling pathways associated with inflammation is increased. Also, fibrosis is the development of excess fibrous tissue in organ or tissue. It is an abnormal phenomenon in which the extent of fibrous connective tissues is increased. In PKD, increases in the activity of molecules such as growth factor and TGF-β have been reported to occur and promote fibrosis. Therefore, the inflammation and fibrosis responses have been suggested as therapeutic targets for PKD. In order to guide further studies, this review indicates the roles of inflammatory and fibrosis signaling in PKD.

Keywords Inflammation • Fibrosis • Polycystic kidney disease • Macrophage • TNF-alpha • TGF-beta

4.1 Introduction

Inflammation is a kind of protective mechanism involved in the response to foreign organisms. When inflammation occurs, it typically shows redness, heat, pain, swelling, and so on. Inflammation is especially characteristic of the innate immune response.

H. Mun
Molecular Medicine Laboratory, Department of Life systems, Sookmyung Women's University, Cheongpa-ro 47-gil 100, Yongsan-gu, Seoul 04310, South Korea
e-mail: nicegoto@sookmyung.ac.kr

J.H. Park (✉)
Department of Life systems, Sookmyung Women's University, Cheongpa-ro 47-gil 100, Yongsan-gu, Seoul 04310, South Korea
e-mail: parkjh@sookmyung.ac.kr

J.H. Park, C. Ahn (eds.), *Cystogenesis*, Advances in Experimental Medicine and Biology 933, DOI 10.1007/978-981-10-2041-4_4

Fibrosis is characterized by the excessive production and accumulation of collagen, accompanied by the decomposition of connective tissues (Birbrair et al. 2014) and is a central feature of several diseases (Zeisberg and Kalluri 2013). Fibrosis responds to injury and also mediates scar formation.

In general, increased inflammation and fibrosis aggravates autosomal polycystic kidney disease (ADPKD). While these processes are not thought to be primary causes of ADPKD, some cells and molecules related to inflammation and/or fibrosis can influence renal function and the progression of ADPKD. For example, macrophages, pro-inflammatory chemoattractants such as monocyte chemoattractant protein 1 (MCP-1), and chemokines such as tumor necrosis factor (TNF)-α, interleukin (IL)-1, and transforming growth factor (TGF)-β can contribute to accelerating disease progression.

4.2 Inflammation and Polycystic Kidney Disease (PKD)

In PKD, an accumulation of macrophages accelerates the proliferation of cystic lining cells and increases the abundance of macrophage-associated factors such as Arg1, Ccl (chemokine ligand), and IL-10. The deletion of macrophages via treatment with liposomal clodronate has been observed to decrease the proliferation of tubular cells and cystic-lining cells and improve renal function and disease progression (Karihaloo et al. 2011). Macrophages induce the production of IL-1β by stimulating the release of prostaglandin E2 (Sugimoto and Narumiya 2007).

M1-like macrophages produce TNF-α, thus encouraging inflammation, and M2-like macrophages stimulate the proliferation of tubular cells and fibrosis formation, thus playing a role in tissue repair. In normal or PKD conditions, renal epithelial cells promote the differentiation of naïve macrophages to M2-like macrophages. The proportion of cells expressing CD163, the marker of M2-like macrophages, has been reported to be increased in ADPKD and autosomal recessive polycystic kidney disease (ARPKD) (Swenson-Fields et al. 2013).

When the expression of the *PKD1* gene decreases, the levels of macrophage chemoattractants such as MCP1 and CXCL16 increases and promotes macrophage migration (Karihaloo et al. 2011). PCL-null cells form cysts and elevate the secretion of MCP1 and CXCL16 (Karihaloo et al. 2011). Also, MCP1 exists in the urine and cystic fluid of ADPKD patients (Zheng et al. 2003) and promotes the growth of cysts, stimulating increases in the levels of ED-1 positive macrophages and the mRNA levels of *MCP1* (Cowley et al. 2001). As a result, the production of MCP1 further increases, exacerbating declines in kidney function and promoting disease progression.

Macrophage migration inhibitory factor (MIF) is up-regulated in mouse and human ADPKD (Chen et al. 2015). Increased production of MIF has been shown in polycystin-1-deficient murine kidney and ADPKD patients' cyst fluid. MIF regulates various cell proliferation pathways through extracellular signal-regulated kinase (ERK), mammalian target of rapamycin (mTOR), and the Rb/E2F pathway, as well as glucose uptake, ATP production, and apoptosis via p53-dependent signaling. In addition, MIF induces the recruitment of macrophages, which increase the secretion of MCP1 and pro-inflammatory cytokines including TNF-α, IL-1β, and IL-2 (Gregory et al. 2006). Especially, MIF and TNF-α form a positive feedback loop as MIF increases the recruitment of macrophages that secrete TNF-α and the increased amount of TNF-α stimulates the expression of MIF (Chen et al. 2015).

Cysts release high levels of pro-inflammatory cytokines, for example IL-1β, TNF-α, and IL-2 (Gardner et al. 1991). TNF-α is a cytokine that is related to the inflammatory response and is known to be regulated by macrophages. TNF-α can bind to two types of receptor. TNF receptor (TNFR)-1 is expressed in most tissues, while TNFR-2 is expressed only in cells of the immune system (Wajant et al. 2003). TNF-α converting enzyme (TACE) cleaves membrane-bound TNF (mTNF) to soluble TNF (sTNF). mTNF and sTNF bind with TNFR and activate various signaling pathways; for example, the NF-kB, MAPK, and JNK pathways.

TNF-α is present in the cystic fluid of human PKD patients along with TNF-α converting enzyme (TACE) and TNFR-I and II. TNF-α increases the expression of TACE, FIP, and TNFR-I (Li et al. 2008), though TNFR-I also positively regulates TNF-α activity (De Groote et al. 1993). Also, TNF-α plays a role in the activation of the mTOR pathway through the suppression of TSC1 (Lee et al. 2007). Also, related to its effect of inducing macrophage infiltration, TNF-α activates the Akt/mTOR and MAPK/CDK2 pathways through the induction of ID2 protein (Zhou et al. 2015) and TNF-α inhibits the interaction between PC1 and PC2 (Li et al. 2008; Ta et al. 2013). As a result, the accumulation of TNF-α occurs and the size of the cyst and disease progression are aggravated (Li et al. 2008).

Interstitial inflammation is related to the nuclear factor (NF)-kB and JAK-STAT pathways, which are correlated with PKD. NF-kB is a transcription factor that is involved in cell survival and the immune response. In canonical NF-kB signaling, when NF-kB is stimulated by factors such as TNF-α, IL-2, or lymphocytes, activated IKK complex phosphorylates IkB proteins, leading to polyubiquitination. Ubiquitinated IkB is degraded by the proteasome, freeing the p50/p65 NF-kB dimer to translocate to the nucleus and activate gene transcription (Skaug et al. 2009; Gilmore 2006; Jost and Ruland 2007).

NF-kB regulates pro-inflammatory cytokines including TNF-α, IL-1α, CCL3, CCL4, and MCP1 (Tak and Firestein 2001). Although polycystin-1 elevates NF-kB activation, leading to protection against apoptosis and the promotion of cell proliferation (Banzi et al. 2006), most studies have indicated that, in *PKD1*$^{-/-}$ cells, NF-kB activity and the phosphorylation of p65 are activated through pro-inflammatory mediators (Qin et al. 2012). Also, FC1 (fibrocystin/polyduction) depleted cells of ARPKD also show NF-kB activation (Mangolini et al. 2010). According to these studies, NF-kB is increased in PKD and induces the inflammation response. The inhibition of NF-kB is thought to represent a therapeutic strategy for PKD; however, inhibiting NF-kB could promote apoptosis (Kaltschmidt et al. 2000) and thereby alleviate PKD progression. Therefore, the role of NF-kB in PKD requires further study.

The Janus kinase (JAK) and signal transducer and activator of transcription (STAT) pathway influences cell proliferation, differentiation, transcription, and immune responses (Rawlings et al. 2004). When ligands bind to the JAK receptors and cause them to form dimers, the tyrosine kinase activity of JAK is activated. The activation of JAK phosphorylates STAT proteins, after which STAT assembles into the dimer form. Dimer STATs can perform a role as transcription factors (Aaronson and Horvath 2002).

Malfunction of the JAK-STAT pathway causes inflammation in PKD. PC1 and PC2 are essential for the regulation of JAK1, JAK2, and STATs (Bhunia et al. 2002). In *PKD1*$^{-/-}$ mouse embryos, increased phosphorylation of STAT1 and p21waf1 activation were observed (Bhunia et al. 2002). Inflammation in PKD via JAK-STAT pathway failure is increased by PKD1 and PKD2 mutations. In the PKD condition, the soluble tail of PC1 modulates the activities of both STAT1 and STAT3 through interactions with cytokines and JAK, and abnormal STAT3 activity promotes proliferation (Talbot et al. 2011). STAT3 is regulated by soluble PC1 in a Src-dependent manner, by the inhibition of suppressor of cytokine signaling 3 (Talbot et al. 2014). Also, PC1 interacts with STAT6, leading to increases in STAT6-dependent gene expression (Low et al. 2006). STAT6 also accumulates in cystic-lining cells (Low et al. 2006). STAT6 is up-regulated in cystic-lining cells by the activation of the IL-13 and IL-4/13 receptors, and suppressor of STAT6 reduces cell proliferation and growth as well as disease progression (Olsan et al. 2011). Therefore, the inhibition of STAT3 and STAT6 may be a therapeutic target of ADPKD (Brosius and He 2015). Overall, when inflammation is up-regulated because of PKD, the secretion of cytokines stimulates the JAK-STAT signaling pathway. Then, the activated JAK-STAT signaling pathway promotes proliferation, apoptosis, and inflammation.

4.3 Fibrosis and PKD

When an imbalance between growth factors and cytokines is generated in PKD, the abundance of pro-fibrotic factors such as TGF-β1, connective tissue growth factor, platelet-derived growth factor, fibroblast growth factor-2, and osteopontin are increased, whereas the abundance of anti-fibrotic factors such as hepatocyte growth factor and bone morphogenic protein-7 is decreased (Lan 2011; Grantham 1997; Norman 2011).

As TGF-β is a kind of cytokine, the TGF-β pathway is usually related to proliferation, apoptosis, cell-cell interaction, and cell differentiation. TGF-β is secreted by macrophages, dendritic cells, and lymphocytes. However, TGF-β also functions as a pro-fibrotic cytokine (Sureshbabu et al. 2016) and activates SMAD signaling (Vernon et al. 2010). When TGF-β binds to its receptors, TGF-β receptor types I and II undergo assembly and activate receptor type I. Then, the activated receptor phosphorylates receptor-regulated SMAD (R-SMAD), which binds to the common mediator SMAD (co-SMAD). This complex can translocate to the nucleus to activate gene transcription (Massague 2000).

The abundance of TGF-β is increased in cystic-lining cells (Klingel et al. 1991), which causes fibrogenesis in end-stage renal failure and decreases the abundance of epithelial-mesenchymal transition (EMT) markers (Chea and Lee 2009). TGF-β-SMAD may not have a major effect on cyst formation in the early stages of renal dysfunction. In addition, the expression of TGFR1 and 2 was found to be elevated in the *PKD1*$^{L3/L3}$ mouse model (Chen et al. 2008). Therefore, the inhibition of TGF-β could decrease cyst formation and progression (Chea and Lee 2009). However, TGF-β2 alleviates disease progression and cystogenesis by controlling the synthesis of extracellular matrix (ECM) proteins and cell adhesion (Elberg et al. 2012).

The ECM is a complex network of proteins that fills the extracellular space. It is involved in connective tissues and includes various components; for example, proteoglycan, collagen, elastin, and fibronectin. Its main functions are supporting cells, cell-cell adhesion, differentiation, proliferation, and fibrosis.

Matrix metalloproteinases (MMPs) are able to degrade ECM proteins. Several classes of MMPs have been characterized, including collagenases such as MMP1 and MMP8, gelatinases, membrane-type MMPs, metalloelastase, and others. MMPs are inhibited by tissue inhibitors of metalloproteinases (TIMPs). Maintaining an appropriate balance between MMPs and TIMPs is important for normal homeostasis of the ECM. When imbalances between MMPs and TIMPs occur, this can trigger diseases such as cancer, arthritis, and fibrosis.

In the kidneys, several types of MMPs and TIMPs are expressed, including MMP-2, 3, 9, 13, 14, 24, 25, 27, and 28 and TIMP-1, 2, and 3 (Catania et al. 2007). Especially, MMP-1, 2, 9, and 14 and TIMP-1 are related to PKD. Increased collagen expression and MMPs activity stimulate the formation of cysts in PKD (Liu et al. 2012). MMP2 is down-regulated in PKD in the Han:SPRD rat model, along with the up-regulation of TIMP-1 and TIMP-2 (Schaefer et al. 1996). Some MMPs and TIMPs, such as MMP-2, 3, and 9 and TIMP-1 and 2, are up-regulated in cystic cells compared with normal cells (Rankin et al. 1996). The increased expression of MMPs and TIMPs induces remodeling and thickening of the cystic membrane. As a result, fibrosis is induced in cystic-lining cells. Therefore, the inhibition of MMP-2, MMP-14, and TIMP-2 by sirolimus decreases the accumulation of ECM and alleviates cystic kidney disease (Berthier et al. 2008). Similarly, the suppression of MMP-14 though batimastat was reported to reduce cyst formation and kidney weight (Gardner et al. 1991).

There are three types of Wnt signaling pathways, the canonical Wnt pathway, the non-canonical planar cell polarity pathway, and the non-canonical Wnt/Ca^{2+} pathway (Komiya and Habas 2008). Wnt signaling regulates gene transcription, cytoskeletal structure, proliferation, and cell migration. In the canonical Wnt signaling pathway, when Wnt is absent, complexes that contain disheveled, axin, glycogen synthase kinase-3β, and adenomatous polyposis coli degrade β-catenin (Logan and Nusse 2004). On the other hand, when Wnt is present, that complex is inhibited, and then β-catenin is accumulated and functions as a transcription factor.

Because β-catenin regulates the process of EMT, the Wnt signaling pathways are closely connected with fibrosis (Kim et al. 2002). It has been reported that a trans-

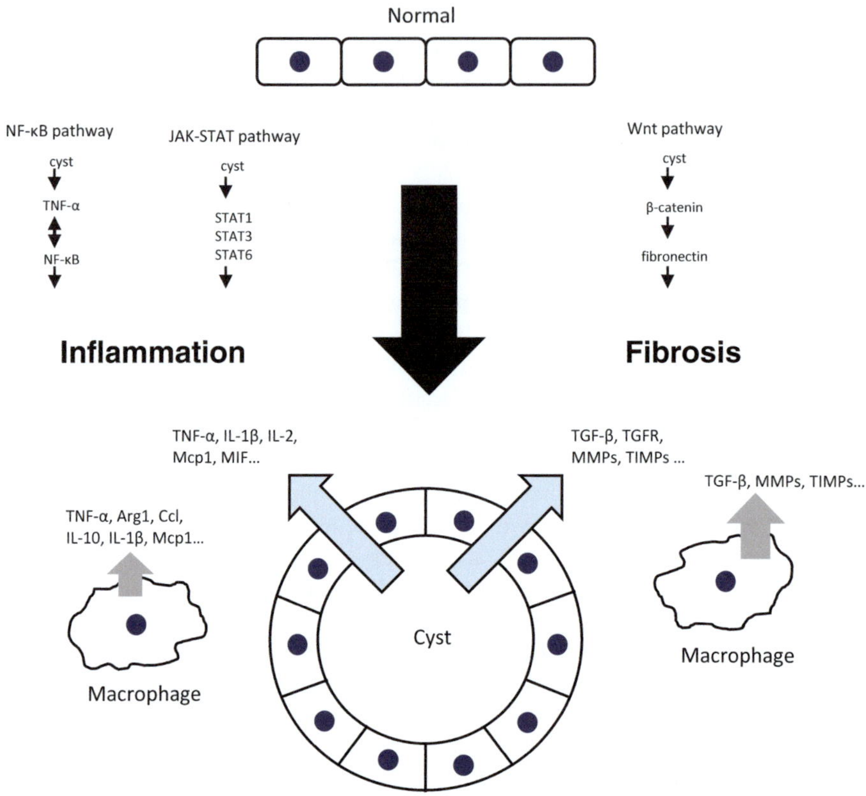

Fig. 4.1 A schematic summary of inflammation and fibrosis in ADPKD

genic mouse model that constitutively expresses β-catenin develops PKD (Saadi-Kheddouci et al. 2001), and the activity of Wnt signaling in cystic-lining cells is increased in PKD patients (Lal et al. 2008). The expression of Wnt4, which is particularly related to the EMT process (Stark et al. 1994), is elevated in jck mice that show PKD (Surendran et al. 2002). In *Gpr48*$^{-/-}$ mice that develop PKD, renal fibrosis was shown to accompany the activation of the Wnt signaling pathway (Dang et al. 2014). Also, Wnt signaling exerts effects on primary cilia and regulates cilia formation (Cisternas et al. 2014). Transgenic mice with genetic knock outs of the ciliary proteins Kif3a and Bbs1 showed a hyperactivation of Wnt signaling compared with normal mice (Corbit et al. 2008). Likewise, the up-regulation of Wnt/β--catenin signaling induces an increased expression of the gene encoding fibronectin, which is closely related to fibrosis (Cisternas et al. 2014). Therefore, the excessive activation of the Wnt signaling pathway results in an increased frequency of EMT (Kim et al. 2002), leading to fibrosis (Fig. 4.1).

4.4 Conclusion

In PKD, it has been shown that the inflammation response and fibrosis are increased in end-stage renal disease. Inflammation can initiate fibrosis, causing thickening and remodeling of the cystic-lining cell membrane. However, inflammation and fibrosis alone are not able to generate PKD, which also requires hyper-proliferation to occur in the kidneys.

Following an increase in cell proliferation, the abundance of certain cells and cytokines is elevated. Macrophage and cytokines secreted by macrophages accumulate in cysts and urine (Swenson-Fields et al. 2013; Zheng et al. 2003; Cowley et al. 2001; Chen et al. 2015; Gregory et al. 2006). In addition, pro-inflammatory cytokines are up-regulated in PKD. It was reported that an increased concentration of TNF-α induces inflammation and proliferation (Li et al. 2008; Ta et al. 2013). TGF-β, MMPs, and TIMPs trigger fibrosis, especially TGF-β, which is involved in fibrosis as an inflammatory cytokine. MMPs and TIMPs induce the re-structuring of cell formations and the accumulation of collagen and fibroblasts.

An increase in the abundance of cytokines and growth factors can lead to the development of inflammation and fibrosis. Moreover, these secreted factors can also induce proliferation. Thus, the induction of inflammation and fibrosis by hyper-proliferation further accelerates cell proliferation, resulting in a vicious cycle among proliferation, inflammation, and fibrosis.

Some reports have shown that the inhibition of cytokines that are involved in the up-regulation of inflammation and fibrosis could be a therapeutic strategy for PKD (Kaltschmidt et al. 2000; Brosius and He 2015; Chea and Lee 2009; Berthier et al. 2008; Obermuller et al. 2001). However, it has been reported that the inhibition of cytokines can further exacerbate certain degenerative diseases (Kaltschmidt et al. 2000; Elberg et al. 2012), possibly due to them playing a pro-apoptotic role. Because cytokines have functions not only in the immune response but are also connected with various signaling pathways. Therefore, the effects of regulating cytokines using therapeutic agents must be carefully determined through further research.

References

Aaronson DS, Horvath CM (2002) A road map for those who don't know JAK-STAT. Science 296(5573):1653–1655. doi:10.1126/science.1071545

Banzi M, Aguiari G, Trimi V, Mangolini A, Pinton P, Witzgall R, Rizzuto R, del Senno L (2006) Polycystin-1 promotes PKCalpha-mediated NF-kappaB activation in kidney cells. Biochem Biophys Res Commun 350(2):257–262. doi:10.1016/j.bbrc.2006.09.042

Berthier CC, Wahl PR, Le Hir M, Marti HP, Wagner U, Rehrauer H, Wuthrich RP, Serra AL (2008) Sirolimus ameliorates the enhanced expression of metalloproteinases in a rat model of autosomal dominant polycystic kidney disease. Nephrol Dial Transplant Off Publ Eur Dial Trans Assoc Eur Renal Assoc 23(3):880–889. doi:10.1093/ndt/gfm697

Bhunia AK, Piontek K, Boletta A, Liu L, Qian F, Xu PN, Germino FJ, Germino GG (2002) PKD1 induces p21(waf1) and regulation of the cell cycle via direct activation of the JAK-STAT signaling pathway in a process requiring PKD2. Cell 109(2):157–168

Birbrair A, Zhang T, Files DC, Mannava S, Smith T, Wang ZM, Messi ML, Mintz A, Delbono O (2014) Type-1 pericytes accumulate after tissue injury and produce collagen in an organ-dependent manner. Stem Cell Res Ther 5(6):122. doi:10.1186/scrt512

Brosius FC 3rd, He JC (2015) JAK inhibition and progressive kidney disease. Curr Opin Nephrol Hypertens 24(1):88–95. doi:10.1097/MNH.0000000000000079

Catania JM, Chen G, Parrish AR (2007) Role of matrix metalloproteinases in renal pathophysiologies. Am J Physiol Renal Physiol 292(3):F905–F911. doi:10.1152/ajprenal.00421.2006

Chea SW, Lee KB (2009) TGF-beta mediated epithelial-mesenchymal transition in autosomal dominant polycystic kidney disease. Yonsei Med J 50(1):105–111. doi:10.3349/ymj.2009.50.1.105

Chen WC, Tzeng YS, Li H (2008) Gene expression in early and progression phases of autosomal dominant polycystic kidney disease. BMC Res Notes 1:131. doi:10.1186/1756-0500-1-131

Chen L, Zhou X, Fan LX, Yao Y, Swenson-Fields KI, Gadjeva M, Wallace DP, Peters DJ, Yu A, Grantham JJ, Li X (2015) Macrophage migration inhibitory factor promotes cyst growth in polycystic kidney disease. J Clin Invest 125(6):2399–2412. doi:10.1172/JCI80467

Cisternas P, Vio CP, Inestrosa NC (2014) Role of Wnt signaling in tissue fibrosis, lessons from skeletal muscle and kidney. Curr Mol Med 14(4):510–522

Corbit KC, Shyer AE, Dowdle WE, Gaulden J, Singla V, Chen MH, Chuang PT, Reiter JF (2008) Kif3a constrains beta-catenin-dependent Wnt signalling through dual ciliary and non-ciliary mechanisms. Nat Cell Biol 10(1):70–76. doi:10.1038/ncb1670

Cowley BD Jr, Ricardo SD, Nagao S, Diamond JR (2001) Increased renal expression of monocyte chemoattractant protein-1 and osteopontin in ADPKD in rats. Kidney Int 60(6):2087–2096. doi:10.1046/j.1523-1755.2001.00065.x

Dang Y, Liu B, Xu P, Zhu P, Zhai Y, Liu M, Ye X (2014) Gpr48 deficiency induces polycystic kidney lesions and renal fibrosis in mice by activating Wnt signal pathway. PLoS One 9(3):e89835. doi:10.1371/journal.pone.0089835

De Groote D, Grau GE, Dehart I, Franchimont P (1993) Stabilisation of functional tumor necrosis factor-alpha by its soluble TNF receptors. Eur Cytokine Netw 4(5):359–362

Elberg D, Jayaraman S, Turman MA, Elberg G (2012) Transforming growth factor-beta inhibits cystogenesis in human autosomal dominant polycystic kidney epithelial cells. Exp Cell Res 318(13):1508–1516. doi:10.1016/j.yexcr.2012.03.021

Gardner KD Jr, Burnside JS, Elzinga LW, Locksley RM (1991) Cytokines in fluids from polycystic kidneys. Kidney Int 39(4):718–724

Gilmore TD (2006) Introduction to NF-kappaB: players, pathways, perspectives. Oncogene 25(51):6680–6684. doi:10.1038/sj.onc.1209954

Grantham JJ (1997) Mechanisms of progression in autosomal dominant polycystic kidney disease. Kidney Int Suppl 63:S93–S97

Gregory JL, Morand EF, McKeown SJ, Ralph JA, Hall P, Yang YH, McColl SR, Hickey MJ (2006) Macrophage migration inhibitory factor induces macrophage recruitment via CC chemokine ligand 2. J Immunol 177(11):8072–8079

Jost PJ, Ruland J (2007) Aberrant NF-kappaB signaling in lymphoma: mechanisms, consequences, and therapeutic implications. Blood 109(7):2700–2707. doi:10.1182/blood-2006-07-025809

Kaltschmidt B, Kaltschmidt C, Hofmann TG, Hehner SP, Droge W, Schmitz ML (2000) The pro- or anti-apoptotic function of NF-kappaB is determined by the nature of the apoptotic stimulus. Eur J Biochem/FEBS 267(12):3828–3835

Karihaloo A, Koraishy F, Huen SC, Lee Y, Merrick D, Caplan MJ, Somlo S, Cantley LG (2011) Macrophages promote cyst growth in polycystic kidney disease. J Am Soc Nephrol JASN 22(10):1809–1814. doi:10.1681/ASN.2011010084

Kim K, Lu Z, Hay ED (2002) Direct evidence for a role of beta-catenin/LEF-1 signaling pathway in induction of EMT. Cell Biol Int 26(5):463–476

Klingel R, Storkel S, Dippold W, Rumpelt HJ, Moll R, Kohler H, Meyer zum Buschenfelde KH (1991) Autosomal dominant polycystic kidney disease--in vitro culture of cyst-lining epithelial cells. Virchows Arch B Cell Pathol Incl Mol Pathol 61(3):189–199

Komiya Y, Habas R (2008) Wnt signal transduction pathways. Organogenesis 4(2):68–75

Lal M, Song X, Pluznick JL, Di Giovanni V, Merrick DM, Rosenblum ND, Chauvet V, Gottardi CJ, Pei Y, Caplan MJ (2008) Polycystin-1 C-terminal tail associates with beta-catenin and inhibits canonical Wnt signaling. Hum Mol Genet 17(20):3105–3117. doi:10.1093/hmg/ddn208

Lan HY (2011) Diverse roles of TGF-beta/Smads in renal fibrosis and inflammation. Int J Biol Sci 7(7):1056–1067

Lee DF, Kuo HP, Chen CT, Hsu JM, Chou CK, Wei Y, Sun HL, Li LY, Ping B, Huang WC, He X, Hung JY, Lai CC, Ding Q, Su JL, Yang JY, Sahin AA, Hortobagyi GN, Tsai FJ, Tsai CH, Hung MC (2007) IKK beta suppression of TSC1 links inflammation and tumor angiogenesis via the mTOR pathway. Cell 130(3):440–455. doi:10.1016/j.cell.2007.05.058

Li X, Magenheimer BS, Xia S, Johnson T, Wallace DP, Calvet JP, Li R (2008) A tumor necrosis factor-alpha-mediated pathway promoting autosomal dominant polycystic kidney disease. Nat Med 14(8):863–868. doi:10.1038/nm1783

Liu B, Li C, Liu Z, Dai Z, Tao Y (2012) Increasing extracellular matrix collagen level and MMP activity induces cyst development in polycystic kidney disease. BMC Nephrol 13:109. doi:10.1186/1471-2369-13-109

Logan CY, Nusse R (2004) The Wnt signaling pathway in development and disease. Annu Rev Cell Dev Biol 20:781–810. doi:10.1146/annurev.cellbio.20.010403.113126

Low SH, Vasanth S, Larson CH, Mukherjee S, Sharma N, Kinter MT, Kane ME, Obara T, Weimbs T (2006) Polycystin-1, STAT6, and P100 function in a pathway that transduces ciliary mechanosensation and is activated in polycystic kidney disease. Dev Cell 10(1):57–69. doi:10.1016/j.devcel.2005.12.005

Mangolini A, Bogo M, Durante C, Borgatti M, Gambari R, Harris PC, Rizzuto R, Pinton P, Aguiari G, del Senno L (2010) NF-kappaB activation is required for apoptosis in fibrocystin/polyductin-depleted kidney epithelial cells. Apoptosis Int J Program Cell Death 15(1):94–104. doi:10.1007/s10495-009-0426-7

Massague J (2000) How cells read TGF-beta signals. Nat Rev Mol Cell Biol 1(3):169–178. doi:10.1038/35043051

Norman J (2011) Fibrosis and progression of autosomal dominant polycystic kidney disease (ADPKD). Biochim Biophys Acta 1812(10):1327–1336. doi:10.1016/j.bbadis.2011.06.012

Obermuller N, Morente N, Kranzlin B, Gretz N, Witzgall R (2001) A possible role for metalloproteinases in renal cyst development. Am J Physiol Renal Physiol 280(3):F540–F550

Olsan EE, Mukherjee S, Wulkersdorfer B, Shillingford JM, Giovannone AJ, Todorov G, Song X, Pei Y, Weimbs T (2011) Signal transducer and activator of transcription-6 (STAT6) inhibition suppresses renal cyst growth in polycystic kidney disease. Proc Natl Acad Sci U S A 108(44):18067–18072. doi:10.1073/pnas.1111966108

Qin S, Taglienti M, Cai L, Zhou J, Kreidberg JA (2012) c-Met and NF-kappaB-dependent overexpression of Wnt7a and -7b and Pax2 promotes cystogenesis in polycystic kidney disease. J Am Soc Nephrol JASN 23(8):1309–1318. doi:10.1681/ASN.2011030277

Rankin CA, Suzuki K, Itoh Y, Ziemer DM, Grantham JJ, Calvet JP, Nagase H (1996) Matrix metalloproteinases and TIMPS in cultured C57BL/6 J-cpk kidney tubules. Kidney Int 50(3):835–844

Rawlings JS, Rosler KM, Harrison DA (2004) The JAK/STAT signaling pathway. J Cell Sci 117(Pt 8):1281–1283. doi:10.1242/jcs.00963

Saadi-Kheddouci S, Berrebi D, Romagnolo B, Cluzeaud F, Peuchmaur M, Kahn A, Vandewalle A, Perret C (2001) Early development of polycystic kidney disease in transgenic mice expressing an activated mutant of the beta-catenin gene. Oncogene 20(42):5972–5981. doi:10.1038/sj.onc.1204825

Schaefer L, Han X, Gretz N, Hafner C, Meier K, Matzkies F, Schaefer RM (1996) Tubular gelatinase A (MMP-2) and its tissue inhibitors in polycystic kidney disease in the Han:SPRD rat. Kidney Int 49(1):75–81

Skaug B, Jiang X, Chen ZJ (2009) The role of ubiquitin in NF-kappaB regulatory pathways. Annu Rev Biochem 78:769–796. doi:10.1146/annurev.biochem.78.070907.102750

Stark K, Vainio S, Vassileva G, McMahon AP (1994) Epithelial transformation of metanephric mesenchyme in the developing kidney regulated by Wnt-4. Nature 372(6507):679–683. doi:10.1038/372679a0

Sugimoto Y, Narumiya S (2007) Prostaglandin E receptors. J Biol Chem 282(16):11613–11617. doi:10.1074/jbc.R600038200

Surendran K, McCaul SP, Simon TC (2002) A role for Wnt-4 in renal fibrosis. Am J Physiol Renal Physiol 282(3):F431–F441. doi:10.1152/ajprenal.0009.2001

Sureshbabu A, Muhsin SA, Choi ME (2016) TGF-beta signaling in the kidney: pro-fibrotic and protective effects. Am J Physiol Renal Physiol 00365:02015. doi:10.1152/ajprenal.00365.2015

Swenson-Fields KI, Vivian CJ, Salah SM, Peda JD, Davis BM, van Rooijen N, Wallace DP, Fields TA (2013) Macrophages promote polycystic kidney disease progression. Kidney Int 83(5):855–864. doi:10.1038/ki.2012.446

Ta MH, Harris DC, Rangan GK (2013) Role of interstitial inflammation in the pathogenesis of polycystic kidney disease. Nephrology 18(5):317–330. doi:10.1111/nep.12045

Tak PP, Firestein GS (2001) NF-kappaB: a key role in inflammatory diseases. J Clin Invest 107(1):7–11. doi:10.1172/JCI11830

Talbot JJ, Shillingford JM, Vasanth S, Doerr N, Mukherjee S, Kinter MT, Watnick T, Weimbs T (2011) Polycystin-1 regulates STAT activity by a dual mechanism. Proc Natl Acad Sci U S A 108(19):7985–7990. doi:10.1073/pnas.1103816108

Talbot JJ, Song X, Wang X, Rinschen MM, Doerr N, LaRiviere WB, Schermer B, Pei YP, Torres VE, Weimbs T (2014) The cleaved cytoplasmic tail of polycystin-1 regulates Src-dependent STAT3 activation. J Am Soc Nephrol JASN 25(8):1737–1748. doi:10.1681/ASN.2013091026

Vernon MA, Mylonas KJ, Hughes J (2010) Macrophages and renal fibrosis. Semin Nephrol 30(3):302–317. doi:10.1016/j.semnephrol.2010.03.004

Wajant H, Pfizenmaier K, Scheurich P (2003) Tumor necrosis factor signaling. Cell Death Differ 10(1):45–65. doi:10.1038/sj.cdd.4401189

Zeisberg M, Kalluri R (2013) Cellular mechanisms of tissue fibrosis. 1. Common and organ-specific mechanisms associated with tissue fibrosis. Am J Physiol Cell Physiol 304(3):C216–C225. doi:10.1152/ajpcell.00328.2012

Zheng D, Wolfe M, Cowley BD Jr, Wallace DP, Yamaguchi T, Grantham JJ (2003) Urinary excretion of monocyte chemoattractant protein-1 in autosomal dominant polycystic kidney disease. J Am Soc Nephrol JASN 14(10):2588–2595

Zhou JX, Fan LX, Li X, Calvet JP, Li X (2015) TNFalpha signaling regulates cystic epithelial cell proliferation through Akt/mTOR and ERK/MAPK/Cdk2 mediated Id2 signaling. PLoS One 10(6):e0131043. doi:10.1371/journal.pone.0131043

Chapter 5
Functional Study of the Primary Cilia in ADPKD

Je Yeong Ko

Abstract The primary cilium is a microtubule-based organelle that is considered to be a cellular antennae, because proteins related to multiple signaling pathways such as Wnt, PDGFRα, Hh, and mechanosignaling are localized to the membrane of the primary cilium. In the kidney, primary cilia extend from the cell membrane to the lumen of renal tubules to respond to fluidic stress. Recent studies have indicated that the disruption of ciliary proteins including polycystin-1 (PC1), polycystin-2 (PC2), and members of the intraflagellar transport (IFT) family induce the development of polycystic kidney disease (PKD), suggesting that the malformation or absence of primary cilia is a driving force of the onset of PKD. Therefore, in this chapter, the renal cystogenesis mechanism induced by cilia defects and pathogenic ciliary proteins associated with PKD development will be described.

Keywords Cilia • Ciliogenesis • Cystogenesis • Intraflagellar transport

5.1 Primary Cilia

The cilium is a finger-like structure that protrudes from the apical membrane surface of various vertebrate cells. Cilia have been observed to take two forms, motile or immotile, and are conserved in eukaryotes. For a long time, many studies focused on motility of the cilia, because scientists have considered that immotile cilia (now commonly referred to as primary cilia) are evolutionarily degenerated. However, increasing evidence has indicated that primary cilia have a role in regulating various signaling pathways in most mammalian cells and that primary cilia can sense physical and biochemical signals (Singla and Reiter 2006), suggesting that primary cilia may represent a sensory organelle distributed throughout vertebrate cells. Interestingly, immotile cilia are widespread compared to motile cilia in the human

J.Y. Ko (✉)
Molecular Medicine Laboratory, Department of Life systems, Sookmyung Women's University, Cheongpa-ro 47-gil 100, Yongsan-gu, Seoul 04310, South Korea
e-mail: jeyeong@sookmyung.ac.kr

J.H. Park, C. Ahn (eds.), *Cystogenesis*, Advances in Experimental Medicine and Biology 933, DOI 10.1007/978-981-10-2041-4_5

body. Also, it has been reported that structural or functional defects of the primary cilia are closely related to the onset of various human diseases, including developmental disorders (Satir et al. 2010). For these reasons, it is important to study the function of primary cilia to understand various pathological defects in vertebrates.

What is the difference between motile cilia and primary cilia? Primary cilia differ from motile cilia in their structural aspect. Primary cilia are characterized by an axoneme comprised of nine pairs of outer microtubules (9+0 microtubule arrangement). On the other hand, an additional central microtubule pair is observed in motile cilia (Singla and Reiter 2006). Motile cilia consist of an axoneme of nine outer doublet microtubules surrounding a single pair of inner microtubules (9+2 microtubule arrangement). Motile and primary cilia are extended from the basal body of differentiated and quiescent cells (Kobayashi and Dynlacht 2011). The basal body is derived from recruitment of the centrosome to the plasma membrane and transition of the mother centriole; thus, cell cycle regulation is a critical cellular event in the assembly or disassembly of cilia (Kobayashi and Dynlacht 2011). Besides these structural differences, the number of primary cilia per cell is distinguished from motile cilia. Primary cilia occur singly on epithelial cells, whereas most cells having motile cilia have one or more motile cilia (Michaud and Yoder 2006).

In vertebrates, most motile cilia are displayed on the ependymal cells of brain ventricles, the epithelial cells of trachea and lung, and the cells of the oviducts (Michaud and Yoder 2006). These motile cilia are involved in the movement of mucus or the circulation of body fluid, whereas primary cilia function in the sensing of extracellular stimulus and signal transduction. In fact, recent studies have revealed that various proteins including ion channels, receptors, and transporters involved in signal transduction are localized to the ciliary membrane and basal body (Satir et al. 2010). Primary cilia are regarded as sensory organelles throughout various tissues due to this special characteristic compared to motile cilia. Examples of primary cilia in mammalian cells include those on the renal tubules, the endothelium covering the cornea, and bile duct epithelial cells in liver (Ibanez-Tallon et al. 2003), suggesting that primary cilia play a role in phototransduction, mechanosignaling, chemosensory, and osmosensory function (Waters and Beales 2011).

5.1.1 *Ciliogenesis*

Cilia are highly conserved cellular organelles of most mammalian cells. To assemble cilia, several distinct stages collectively known as ciliogenesis are required. Unlike other cellular organelles, primary cilia are a dynamic structure whose assembly depends on cell cycle progression, because the centrioles involved in cell cycle regulation are an essential component in the formation and maturation of basal body of the cilia. For these reasons, proliferating cells must exit the mitotic cycle and enter the G0 stage to free centrioles for axoneme nucleation (Avasthi and Marshall 2012).

Ciliogenesis consists of several ordered steps. First, basal bodies derived from centrosomes are formed and recruited to the apical membrane of the polarized cells, and then docking and fusion with plasma membrane occur to initiate ciliogenesis (Ishikawa and Marshall 2011; Avasthi and Marshall 2012). Finally, the elongation of the ciliary membrane and axoneme continues by the process of intraflagellar transport (IFT) (Avasthi and Marshall 2012). Because all proteins are synthesized in the cytoplasm, systems for the recruitment of proteins related to cilia assembly or functioning at the ciliary tip are required, and this recruitment process is termed IFT (Ishikawa and Marshall 2011).

IFT proteins, which are involved in the bi-directional movement protein complex, are divided into two categories, IFT-A (IFT43, IFT121, IFT122, IFT139, IFT140, IFT144) and IFT-B (IFT20, IFT22, IFT25, IFT27, IFT46, IFT52, IFT56, IFT57, IFT70, IFT74, IFT80, IFT81, IFT88, IFT172), according to their reaction of movement (Taschner et al. 2012). The direction of the transport of IFT-A proteins is from the ciliary tip to the basal body, mediated by the dynein-2 motor protein, so this complex is thought to contribute to retrograde transport, while IFT-B proteins move from the basal body to the ciliary tip, dependent on the action of the kinesin-II motor protein, and this process is called anterograde transport (Cole et al. 1998; Ishikawa and Marshall 2011). Therefore, IFT proteins are recruited to the ciliary tip by kinesin-II and returned to the cell body by dynein-2, leading to the continuous recycling of IFT proteins (Ishikawa and Marshall 2011).

Recent studies demonstrate that the malfunction of most IFT-B complex proteins leads to developmental defects with the shortening or absence of primary cilia in vitro and in vivo (Pazour et al. 2000; Deane et al. 2001; Jonassen et al. 2008; Follit et al. 2006), indicating that these proteins are involved in cilia elongation and the regulation of tissue homeostasis. In contrast, most IFT-A complex proteins do not seem to be essential for cilia assembly (Efimenko et al. 2006; Tran et al. 2008; Tsao and Gorovsky 2008). Although primary cilia are observed in IFT-A complex protein-disrupted models, dysregulated signaling pathways regulated by primary cilia or the mislocalization of ciliary proteins have been observed in these models (Jonassen et al. 2012).

5.1.2 Signaling Pathways Regulated by Primary Cilia

Although primary cilia have no motility, this organelle has essential roles in the developmental stages and genetic diseases in vertebrates (Goetz and Anderson 2010). An accumulation of scientific evidence has demonstrated that signaling molecules related to cell proliferation, differentiation, cell cycle, survival, and autophagy are localized to the membrane of the primary cilium, suggesting that primary cilia are specialized for sensing or transmitting cellular signals. Here, the representative signaling pathways regulated by primary cilia are introduced.

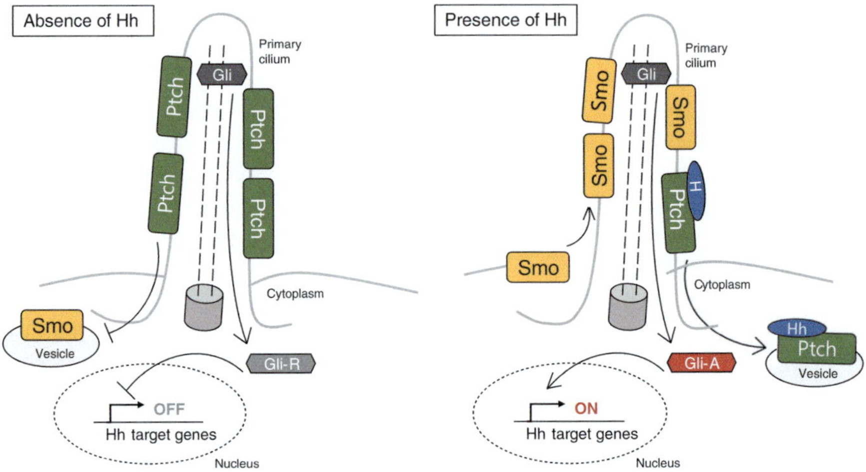

Fig. 5.1 Hh signaling regulated by primary cilia

5.1.2.1 Hedgehog Signaling

Although it has been reported that various signaling pathways are involved in ciliary signaling, the regulatory mechanism of Hedgehog (Hh) signaling in the primary cilia is particularly well understood. Hh signaling is highly conserved in vertebrate development and plays an essential role in regulating tissue patterning and homeostasis (Christensen and Ott 2007; Drummond 2012). Consistent with this, the dysregulation of Hh signaling is observed in models of developmental defects. Recent studies have demonstrated that, in the absence of Hh ligand, Hh ligand binding receptor, Patched (Ptch1), inhibits the recruitment of the Smoothened (Smo) transmembrane protein to the primary cilium (Nozawa et al. 2013) (Fig. 5.1). However, in the presence of Hh ligand, the Hh/Ptch1 protein complex is internalized to the cytosol and Smo proteins accumulate in the ciliary membrane, leading to the expression of Hh target genes via activation of glioma (Gli) transcription factors (Nozawa et al. 2013) (Fig. 5.1).

Several studies have shown that certain ciliary proteins are associated with the localization of Hh signaling molecules. Although mutant mice lacking functional Ift25, which is known as a component of the IFT-B complex, show multiple developmental defects including growth restriction, omphaloceles, and polydactyly, *Ift25* null mice and cells are still ciliated, suggesting that Ift25 is not required for ciliogenesis (Keady et al. 2012). However, interestingly, the same study showed that Hh signaling defects were observed in *Ift25* null mutants and that Ift25 is involved in the regulation of Hh signaling proteins localization on the ciliary membrane (Keady et al. 2012). Other evidence has shown that intestinal cell kinase (Ick), which is known as a ciliary protein, regulates Hh signaling (Moon et al. 2014). *Ick* null mice embryos display an endocrine-cerebro-osteodysplasia (ECO) syndrome-like phenotype with defective Hh signaling (Moon et al. 2014). In addition, this research group

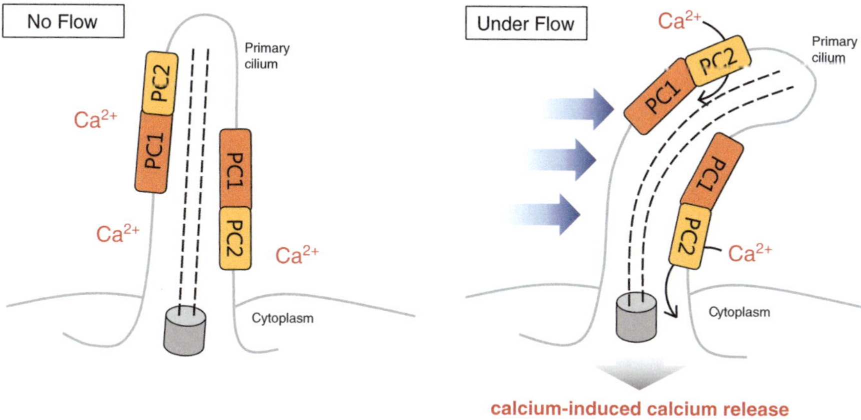

Fig. 5.2 Mechanosignaling regulated by primary cilia

suggested that the knockdown of Ick expression in fibroblasts leads to an increased length of the primary cilia, in which the Smo and Gli2 proteins are abnormally localized (Moon et al. 2014). Taken together, these findings indicate that primary cilia and cilia-related proteins are key regulators of ciliary Hh signaling.

5.1.2.2 Mechanosignaling

Chemical and kinetic stimuli could be detected by the primary cilia, because they are exposed to various hormones, growth factors, chemical substances, fluidic flow, and pressure in the extracellular environment (Basten and Giles 2013). Among the various ciliary responses to extracellular stimuli, in this section, the mechanisms of mechanosignaling by primary cilia are introduced.

The mechanical response of the primary cilia is well understood in renal biology. Representative mechanical stresses include touch, pressure, flow, and vibration, and their detection is collectively referred to as mechanosensation (Basten and Giles 2013). In the renal tubules, which are composed of renal epithelial cells that have primary cilia, the mechanical response to physical phenomena including urine flow occurs through the lumen of the renal tubules. When liquid flow passes through the lumens of the renal tubules, the primary cilia of the renal epithelial cells bend, leading to the initiation of a mechanical response (Fig. 5.2). The deflection of primary cilia in response to luminal flow is recognized by polycystin-1 (PC1) proteins localized to the membrane of primary cilia on the renal epithelial cells. PC1 acts as mechanosensory protein for luminal flow that transmits mechanical stress from the extracellular environment to polycystin-2 (PC2) protein, which is localized to renal primary cilia and associates with PC1 to form a protein complex that functions as a calcium channel (Huang and Lipschutz 2014) (Fig. 5.2). Activated PC2 proteins induce minimal calcium ion influx, leading to massive intracellular calcium release through calcium-induced calcium release (CICR) (Nauli et al. 2003) (Fig. 5.2). Finally, an increased calcium level of the cytosol regulates various signaling pathways related to proliferation and development.

There is scientific evidence showing that the primary cilia act as mechanosensors via PC1/PC2 protein complex in renal epithelial cells. The research group of Nauli et al. suggested that Pkd1null/null cells and cyst-lining cells derived from human autosomal dominant polycystic kidney disease (ADPKD) kidneys fail to respond to shear stress, although *Pkd1*$^{null/null}$ cells still have primary cilia (Nauli et al. 2006).

These findings suggest that renal primary cilia act as antennae to detect shear stress movement through renal tubules, while the PC1/PC2 protein complex contributes to mechanotransduction signaling mediated by calcium in the renal cilia (Nauli et al. 2003).

5.1.2.3 Mammalian Target of Rapamycin Signaling

The mammalian target of rapamycin (mTOR) signaling pathway plays roles in cell size control, cell growth, and metabolism (Wullschleger et al. 2006). mTOR signaling is driven by two multiprotein complexes, mTOR complex 1 (mTORC1) and mTOR complex 2 (mTORC2) (Wrighton 2010). mTORC1 consists of five protein components: mTOR, Raptor, GβL, PRAS40, and Deptor, while mTORC2 consists of six protein components: mTOR, Rictor, mSIN1, Protor-1, mLST8, and Deptor (Laplante and Sabatini 2009). When mTORC1 is activated by amino acids and growth factors, it phosphorylates p70S6 kinase (S6K) and 4E-BP1 to activate protein synthesis and cell proliferation and regulate cell size (Boehlke et al. 2010). On the other hand, Lkb1 tumor suppressor, which is known as a serine/threonine kinase, inhibits mTORC1-mediated signaling via the cellular energy status sensor, AMP-activated protein kinase (AMPK) (Shaw et al. 2004). mTOR pathways are well studied in cancer biology, while the regulatory mechanism or relationship between mTOR and primary cilia is not completely understood. In addition, although mTOR signaling is highly activated and is thought to be a major disease-causing pathway in PKD, which is considered as a ciliary defect disease (Ibraghimov-Beskrovnaya and Natoli 2011), the precise relationship between primary cilia and mTOR in the kidney has not been fully elucidated.

However, in 2010, the research group of Boehlke and colleagues suggested that the primary cilia regulate cell size through mTOR signaling (Boehlke et al. 2010). According to the findings of this research group, Lkb1 localized to primary cilia regulates flow-dependent mTOR activity and shear stress induces AMPK phosphorylation at the basal body, resulting in the inhibition of mTORC1 activity and a reduction in cell size (Boehlke et al. 2010) (Fig. 5.3). This paper is the first scientific evidence showing that primary cilia act as mechanosensor to modulate mTOR signaling, but further studies are needed to identify the machinery that mediate the recruitment of mTOR pathway regulators to the ciliary or basal body, such as IFT.

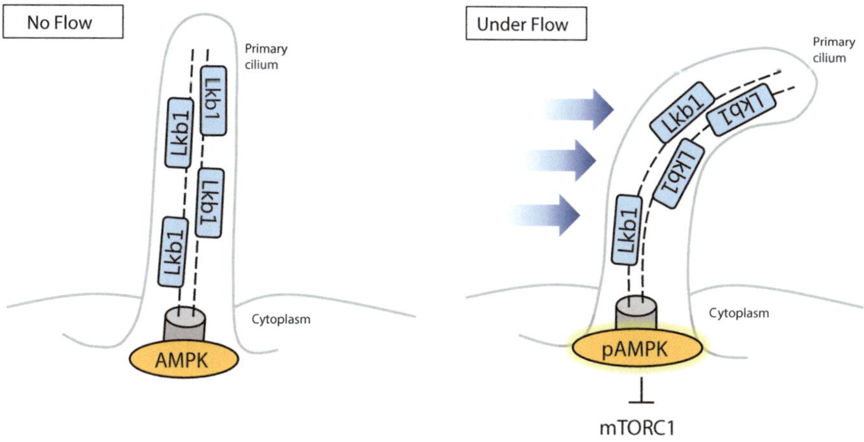

Fig. 5.3 mTOR signaling regulated by primary cilia

5.2 Ciliopathies & PKD

Primary cilia are localized to almost all cells of the human body and function as sensors for the detection of extracellular signals, so it is conceivable that a defective function or structure of the primary cilia is correlated with various human diseases. In fact, defects of the primary cilia are involved in the onset of various diseases, including retinal degeneration, rib/thoracic skeleton defect, pelvic bone defect, polydactyly, liver cysts, hydrocephalus, cardiac defect, mental retardation, airway defect, genital defect, pancreatic cysts, and cystic kidneys (Lee and Gleeson 2011). These multisystemic human diseases are referred to as ciliopathies. Most ciliopathies are caused by mutations of ciliary genes, indicating that the identification of disease-specific functions of mutated ciliary genes is critical for understanding ciliopathies.

Among the many ciliopathies, in this section, we focus on PKD, which is the most common human genetic disorder and is characterized by the formation of fluid-filled cysts in the renal tubules, leading to end-stage renal disease (Patel et al. 2009). Recent studies suggest that PKD is caused by abnormalities of primary cilia, so PKD can be considered as ciliopathy of the kidney. Some studies have demonstrated a relationship between defects of the primary cilia and PKD development. It has been reported that the *PKD1* and *PKD2* genes are mutated in PKD and the proteins encoded by these two genes are localized to the membrane of the renal cilia and form calcium channels, which seem to respond to mechanical stresses including luminal flow through the renal tubules (Chapin and Caplan 2010). Although mutations of these two genes do not affect the ciliogenesis of primary cilia (Ma et al. 2013), a failure to increase calcium influx in response to shear stress is observed in *Pkd1* mutant cells, indicating that defects of polycystin proteins are involved in dysfunctional renal cilia (Nauli et al. 2006; Patel et al. 2009). In addition to these findings, it has been reported that the inactivation of genes related to ciliogenesis including *Ift20*, *Ift88*, and NIMA (never in mitosis A)-related kinase 8 (*Nek8*) cause a severe PKD phenotype in mice (Jonassen et al. 2008; Jonassen et al. 2012; Lehman et al. 2008; Liu et al. 2002). Interestingly, the loss of genes that are required for cilia assembly such as *Ift20* and *Kif3a* induces a renal cystic phenotype in mice with the absence of renal cilia (Jonassen et al. 2008; Lin et al. 2003), whereas one of the PKD mouse models called juvenile cystic kidney (jck) shows a cystic kidney phenotype with lengthened cilia compared to those of normal kidneys (Smith et al. 2006). Taken together, these findings indicate the dysregulated signaling by which defective primary cilia are associated with PKD development and suggest that functional studies of primary cilia and ciliary proteins might contribute to a better understanding of PKD pathogenesis.

5.3 Disruption of Signaling Pathways in Cilia-Defective PKD Mouse Models

The abnormal regulation of signaling pathways such as proliferation, cell cycle, differentiation, apoptosis, inflammation, and fibrosis has been observed in PKD mouse models. Among these pathways, an increase of proliferation has been considered as a major causative factor in the development of PKD. According to a published paper, inactivation of the *Pkd1* gene in mouse kidney cells induces a severe polycystic kidney phenotype with increased proliferation of the cyst-lining epithelial cells (Shibazaki et al. 2008).

Recent studies show that defective primary cilia or the inactivation of cilia-related genes induces aberrant signaling pathways associated with proliferation, differentiation, and development in various PKD mouse models (Ma et al. 2013; Ibraghimov-Beskrovnaya and Natoli 2011; Jonassen et al. 2008; Eguether et al. 2014; Shibazaki et al. 2008). Consistent with this, it has been reported that various signaling pathways including Wingless (Wnt), planar cell polarity, mTOR, mitogen-activated protein kinase (MAPK), and Hh are dysregulated in various ciliopathy models. The major ciliary pathways related to PKD development are discussed in this section.

5.3.1 Increase of MAPK Signaling in PKD with Ciliary Defects

PC1 and PC2 protein complexes have been shown to localize to the ciliary membrane in the kidney. Renal cilia function as a mechanosensor and cooperate with the PC1-PC2 protein complex in the renal tubules to translate mechanical stress through the lumen into an increase of intracellular calcium ion influx (Bastos and Onuchic

2011). When tubular fluid flow induces bending of cilia, the PC2 calcium channel opens to allow calcium ions to enter into the cilioplasm, resulting in CICR from the endoplasmic reticulum (Jin et al. 2014; Winyard and Jenkins 2011). Indeed, calcium entry in response to shear stress is disrupted in renal epithelial cells that have no cilia or cilia with no PC1. These findings demonstrate that the induction of an increase of intracellular calcium in response to fluid flow requires intact primary cilia with PC1 and PC2 protein complexes.

Signaling related to calcium ions is important for maintaining the homeostasis of renal epithelial cells, because most of the calcium-related signaling pathways are associated with cell proliferation in the kidney. Indeed, a lower intracellular calcium level is observed in the renal epithelial cells of PKD (Mangolini et al. 2016; Yamaguchi et al. 2006). This calcium restriction in renal epithelial cells allows the cAMP-dependent activation of MAPK signaling, leading to an increase of cell proliferation, which is a typical hallmark of renal cyst expansion during PKD development (Mangolini et al. 2016).

5.3.2 Increase of mTOR Signaling in PKD with Ciliary Defects

Another aberrant signaling pathway linked to defective primary cilia in PKD is mTOR. Inappropriate activation of mTOR signaling is common feature of PKD and is caused by the inactivation of ciliary genes including *Pkd1* and *Ift88*, suggesting that defects of ciliary function are related to aberrant mTOR signaling (Mostov 2006). Indeed, as described above, it has been revealed that Lkb1, which is known as a negative regulator of mTOR, is localized to primary cilia and inhibits mTORC1 activity under conditions of fluid flow (Boehlke et al. 2010). Consistent with this, the activation of mTOR signaling and an increase of renal epithelial cell size are observed in the kidney of kinesin family member 3A (*Kif3a*)-defective mice models that show a cystic kidney phenotype with a complete loss of renal cilia (Boehlke et al. 2010). Taken together, these findings indicate that primary cilia in the kidney play a role in regulating cell proliferation via the activation of mTOR signaling in response to luminal flow.

5.3.3 Cilia-Dependent Cyst Activating Mechanism in PKD

Primary cilia of the renal epithelial cells seem to function as negative regulators of cyst expansion. As described above, this function is supported by studies on PKD mice models generated by the inactivation of various ciliary genes, indicating that the loss of primary cilia promotes renal cyst formation in vivo. However, in 2013, a newly discovered mechanism referred to as the cilia-dependent cyst activating (CDCA) pathway was proposed. Interestingly, it was suggested that the loss of renal cilia inhibits cyst growth following the loss of polycystins in vivo (Ma et al. 2013).

In addition, it was shown that the size and severity of renal cysts are associated with the length of the time interval between the early loss of polycystins and the subsequent ablation of cilia (Ma et al. 2013). While the loss of cilia or polycystins alone results in the development and progression of renal cysts, renal cilia involution reduces the progression of cyst growth induced by the inactivation of polycystins (Ma et al. 2013; Lee and Somlo 2014). This evidence strongly suggests that the progression of cysts in PKD is regulated by the duration of the interval time between the initial loss of polycystins and the subsequent disappearance of renal cilia (Ma et al. 2013; Lee and Somlo 2014). This is a novel molecular mechanism that explains the relationship between polycystins and primary cilia in the kidney, but the signaling molecules related to this pathway have not yet been elucidated. Therefore, it is critical to define the components related to CDCA to propose answers to the unresolved questions in PKD pathogenesis.

5.4 Concluding Remarks

Primary cilia are thought to be cellular antennae that regulate diverse signaling pathways in mammalian cells. Indeed, various signaling components related to cell proliferation, differentiation, and development are localized to cilia. It has been reported that the dysfunction of renal cilia induced by the inactivation of various ciliary genes is associated with the development of PKD. Emerging evidence has suggested the existence of a primary cilia-independent role of ciliary genes. Recent studies indicated that the protein Ift88, which is known as a critical component of cilia assembly, can also induce cell migration (Boehlke et al. 2015) and is required for the G1/S transition in non-ciliated cells (Robert et al. 2007). These findings indicate that ciliary proteins might have not only cilia-dependent but also cilia-independent roles. In addition, it has been reported that ciliary defect phenotypes differ between tissues, although the same ciliary protein was inactivated in mice (Moon et al. 2014; Chaya et al. 2014), demonstrating that ciliary proteins might have tissue-specific functions. Therefore, defining the cilia-dependent or cilia-independent roles of ciliary proteins and the tissue specificity of ciliary function will lead to a better understanding of the molecular mechanisms of various ciliopathies including PKD.

References

Avasthi P, Marshall WF (2012) Stages of ciliogenesis and regulation of ciliary length. Differ Res Biolog Diver 83(2):S30–S42. doi:10.1016/j.diff.2011.11.015

Basten SG, Giles RH (2013) Functional aspects of primary cilia in signaling, cell cycle and tumorigenesis. Cilia 2(1):6. doi:10.1186/2046-2530-2-6

Bastos AP, Onuchic LF (2011) Molecular and cellular pathogenesis of autosomal dominant polycystic kidney disease. Braz J Med Biol Res = Revista brasileira de pesquisas medicas e biologicas/Sociedade Brasileira de Biofisica [et al] 44(7):606–617

Boehlke C, Kotsis F, Patel V, Braeg S, Voelker H, Bredt S, Beyer T, Janusch H, Hamann C, Godel M, Muller K, Herbst M, Hornung M, Doerken M, Kottgen M, Nitschke R, Igarashi P, Walz G, Kuehn EW (2010) Primary cilia regulate mTORC1 activity and cell size through Lkb1. Nat Cell Biol 12(11):1115–1122. doi:10.1038/ncb2117

Boehlke C, Janusch H, Hamann C, Powelske C, Mergen M, Herbst H, Kotsis F, Nitschke R, Kuehn EW (2015) A cilia independent role of Ift88/polaris during cell migration. PLoS One 10(10):e0140378. doi:10.1371/journal.pone.0140378

Chapin HC, Caplan MJ (2010) The cell biology of polycystic kidney disease. J Cell Biol 191(4):701–710. doi:10.1083/jcb.201006173

Chaya T, Omori Y, Kuwahara R, Furukawa T (2014) ICK is essential for cell type-specific ciliogenesis and the regulation of ciliary transport. EMBO J 33(11):1227–1242. doi:10.1002/embj.201488175

Christensen ST, Ott CM (2007) Cell signaling. A ciliary signaling switch. Science 317(5836):330–331. doi:10.1126/science.1146180

Cole DG, Diener DR, Himelblau AL, Beech PL, Fuster JC, Rosenbaum JL (1998) Chlamydomonas kinesin-II-dependent intraflagellar transport (IFT): IFT particles contain proteins required for ciliary assembly in Caenorhabditis elegans sensory neurons. J Cell Biol 141(4):993–1008

Deane JA, Cole DG, Seeley ES, Diener DR, Rosenbaum JL (2001) Localization of intraflagellar transport protein IFT52 identifies basal body transitional fibers as the docking site for IFT particles. Curr Biol CB 11(20):1586–1590

Drummond IA (2012) Cilia functions in development. Curr Opin Cell Biol 24(1):24–30. doi:10.1016/j.ceb.2011.12.007

Efimenko E, Blacque OE, Ou G, Haycraft CJ, Yoder BK, Scholey JM, Leroux MR, Swoboda P (2006) Caenorhabditis elegans DYF-2, an orthologue of human WDR19, is a component of the intraflagellar transport machinery in sensory cilia. Mol Biol Cell 17(11):4801–4811. doi:10.1091/mbc.E06-04-0260

Eguether T, San Agustin JT, Keady BT, Jonassen JA, Liang Y, Francis R, Tobita K, Johnson CA, Abdelhamed ZA, Lo CW, Pazour GJ (2014) IFT27 links the BBSome to IFT for maintenance of the ciliary signaling compartment. Dev Cell 31(3):279–290. doi:10.1016/j.devcel.2014.09.011

Follit JA, Tuft RA, Fogarty KE, Pazour GJ (2006) The intraflagellar transport protein IFT20 is associated with the Golgi complex and is required for cilia assembly. Mol Biol Cell 17(9):3781–3792. doi:10.1091/mbc.E06-02-0133

Goetz SC, Anderson KV (2010) The primary cilium: a signalling centre during vertebrate development. Nat Rev Genet 11(5):331–344. doi:10.1038/nrg2774

Huang L, Lipschutz JH (2014) Cilia and polycystic kidney disease, kith and kin. Birth Defects Res C Embryo Today Rev 102(2):174–185. doi:10.1002/bdrc.21066

Ibanez-Tallon I, Heintz N, Omran H (2003) To beat or not to beat: roles of cilia in development and disease. Hum Mol Genet 12(Spec No 1):R27–R35

Ibraghimov-Beskrovnaya O, Natoli TA (2011) mTOR signaling in polycystic kidney disease. Trends Mol Med 17(11):625–633. doi:10.1016/j.molmed.2011.06.003

Ishikawa H, Marshall WF (2011) Ciliogenesis: building the cell's antenna. Nat Rev Mol Cell Biol 12(4):222–234. doi:10.1038/nrm3085

Jin X, Mohieldin AM, Muntean BS, Green JA, Shah JV, Mykytyn K, Nauli SM (2014) Cilioplasm is a cellular compartment for calcium signaling in response to mechanical and chemical stimuli. Cell Mol Life Sci CMLS 71(11):2165–2178. doi:10.1007/s00018-013-1483-1

Jonassen JA, San Agustin J, Follit JA, Pazour GJ (2008) Deletion of IFT20 in the mouse kidney causes misorientation of the mitotic spindle and cystic kidney disease. J Cell Biol 183(3):377–384. doi:10.1083/jcb.200808137

Jonassen JA, SanAgustin J, Baker SP, Pazour GJ (2012) Disruption of IFT complex A causes cystic kidneys without mitotic spindle misorientation. J Am Soc Nephrol JASN 23(4):641–651. doi:10.1681/ASN.2011080829

Keady BT, Samtani R, Tobita K, Tsuchya M, San Agustin JT, Follit JA, Jonassen JA, Subramanian R, Lo CW, Pazour GJ (2012) IFT25 links the signal-dependent movement of Hedgehog components to intraflagellar transport. Dev Cell 22(5):940–951. doi:10.1016/j.devcel.2012.04.009

Kobayashi T, Dynlacht BD (2011) Regulating the transition from centriole to basal body. J Cell Biol 193(3):435–444. doi:10.1083/jcb.201101005

Laplante M, Sabatini DM (2009) mTOR signaling at a glance. J Cell Sci 122(Pt 20):3589–3594. doi:10.1242/jcs.051011

Lee JE, Gleeson JG (2011) A systems-biology approach to understanding the ciliopathy disorders. Genome Med 3(9):59. doi:10.1186/gm275

Lee SH, Somlo S (2014) Cyst growth, polycystins, and primary cilia in autosomal dominant polycystic kidney disease. Kidney Res Clin Pract 33(2):73–78. doi:10.1016/j.krcp.2014.05.002

Lehman JM, Michaud EJ, Schoeb TR, Aydin-Son Y, Miller M, Yoder BK (2008) The Oak Ridge Polycystic Kidney mouse: modeling ciliopathies of mice and men. Dev Dyn Off Publ Am Assoc Anat 237(8):1960–1971. doi:10.1002/dvdy.21515

Lin F, Hiesberger T, Cordes K, Sinclair AM, Goldstein LS, Somlo S, Igarashi P (2003) Kidney-specific inactivation of the KIF3A subunit of kinesin-II inhibits renal ciliogenesis and produces polycystic kidney disease. Proc Natl Acad Sci U S A 100(9):5286–5291. doi:10.1073/pnas.0836980100

Liu S, Lu W, Obara T, Kuida S, Lehoczky J, Dewar K, Drummond IA, Beier DR (2002) A defect in a novel Nek-family kinase causes cystic kidney disease in the mouse and in zebrafish. Development 129(24):5839–5846

Ma M, Tian X, Igarashi P, Pazour GJ, Somlo S (2013) Loss of cilia suppresses cyst growth in genetic models of autosomal dominant polycystic kidney disease. Nat Genet 45(9):1004–1012. doi:10.1038/ng.2715

Mangolini A, de Stephanis L, Aguiari G (2016) Role of calcium in polycystic kidney disease: from signaling to pathology. World J Nephrol 5(1):76–83. doi:10.5527/wjn.v5.i1.76

Michaud EJ, Yoder BK (2006) The primary cilium in cell signaling and cancer. Cancer Res 66(13):6463–6467. doi:10.1158/0008-5472.CAN-06-0462

Moon H, Song J, Shin JO, Lee H, Kim HK, Eggenschwiller JT, Bok J, Ko HW (2014) Intestinal cell kinase, a protein associated with endocrine-cerebro-osteodysplasia syndrome, is a key regulator of cilia length and Hedgehog signaling. Proc Natl Acad Sci U S A 111(23):8541–8546. doi:10.1073/pnas.1323161111

Mostov KE (2006) mTOR is out of control in polycystic kidney disease. Proc Natl Acad Sci U S A 103(14):5247–5248. doi:10.1073/pnas.0601352103

Nauli SM, Alenghat FJ, Luo Y, Williams E, Vassilev P, Li X, Elia AE, Lu W, Brown EM, Quinn SJ, Ingber DE, Zhou J (2003) Polycystins 1 and 2 mediate mechanosensation in the primary cilium of kidney cells. Nat Genet 33(2):129–137. doi:10.1038/ng1076

Nauli SM, Rossetti S, Kolb RJ, Alenghat FJ, Consugar MB, Harris PC, Ingber DE, Loghman-Adham M, Zhou J (2006) Loss of polycystin-1 in human cyst-lining epithelia leads to ciliary dysfunction. J Am Soc Nephrol JASN 17(4):1015–1025. doi:10.1681/ASN.2005080830

Nozawa YI, Lin C, Chuang PT (2013) Hedgehog signaling from the primary cilium to the nucleus: an emerging picture of ciliary localization, trafficking and transduction. Curr Opin Genet Dev 23(4):429–437. doi:10.1016/j.gde.2013.04.008

Patel V, Chowdhury R, Igarashi P (2009) Advances in the pathogenesis and treatment of polycystic kidney disease. Curr Opin Nephrol Hypertens 18(2):99–106. doi:10.1097/MNH.0b013e3283262ab0

Pazour GJ, Dickert BL, Vucica Y, Seeley ES, Rosenbaum JL, Witman GB, Cole DG (2000) Chlamydomonas IFT88 and its mouse homologue, polycystic kidney disease gene tg737, are required for assembly of cilia and flagella. J Cell Biol 151(3):709–718

Robert A, Margall-Ducos G, Guidotti JE, Bregerie O, Celati C, Brechot C, Desdouets C (2007) The intraflagellar transport component IFT88/polaris is a centrosomal protein regulating G1-S transition in non-ciliated cells. J Cell Sci 120(Pt 4):628–637. doi:10.1242/jcs.03366
Satir P, Pedersen LB, Christensen ST (2010) The primary cilium at a glance. J Cell Sci 123(Pt 4):499–503. doi:10.1242/jcs.050377
Shaw RJ, Bardeesy N, Manning BD, Lopez L, Kosmatka M, DePinho RA, Cantley LC (2004) The LKB1 tumor suppressor negatively regulates mTOR signaling. Cancer Cell 6(1):91–99. doi:10.1016/j.ccr.2004.06.007
Shibazaki S, Yu Z, Nishio S, Tian X, Thomson RB, Mitobe M, Louvi A, Velazquez H, Ishibe S, Cantley LG, Igarashi P, Somlo S (2008) Cyst formation and activation of the extracellular regulated kinase pathway after kidney specific inactivation of Pkd1. Hum Mol Genet 17(11):1505–1516. doi:10.1093/hmg/ddn039
Singla V, Reiter JF (2006) The primary cilium as the cell's antenna: signaling at a sensory organelle. Science 313(5787):629–633. doi:10.1126/science.1124534
Smith LA, Bukanov NO, Husson H, Russo RJ, Barry TC, Taylor AL, Beier DR, Ibraghimov-Beskrovnaya O (2006) Development of polycystic kidney disease in juvenile cystic kidney mice: insights into pathogenesis, ciliary abnormalities, and common features with human disease. J Am Soc Nephrol JASN 17(10):2821–2831. doi:10.1681/ASN.2006020136
Taschner M, Bhogaraju S, Lorentzen E (2012) Architecture and function of IFT complex proteins in ciliogenesis. Differ Res Biolog Diver 83(2):S12–S22. doi:10.1016/j.diff.2011.11.001
Tran PV, Haycraft CJ, Besschetnova TY, Turbe-Doan A, Stottmann RW, Herron BJ, Chesebro AL, Qiu H, Scherz PJ, Shah JV, Yoder BK, Beier DR (2008) THM1 negatively modulates mouse sonic hedgehog signal transduction and affects retrograde intraflagellar transport in cilia. Nat Genet 40(4):403–410. doi:10.1038/ng.105
Tsao CC, Gorovsky MA (2008) Tetrahymena IFT122A is not essential for cilia assembly but plays a role in returning IFT proteins from the ciliary tip to the cell body. J Cell Sci 121(Pt 4):428–436. doi:10.1242/jcs.015826
Waters AM, Beales PL (2011) Ciliopathies: an expanding disease spectrum. Pediatr Nephrol 26(7):1039–1056. doi:10.1007/s00467-010-1731-7
Winyard P, Jenkins D (2011) Putative roles of cilia in polycystic kidney disease. Biochim Biophys Acta 1812(10):1256–1262. doi:10.1016/j.bbadis.2011.04.012
Wrighton KH (2010) Cell signalling: cilia downsize mTORC1. Nat Rev Mol Cell Biol 11(12):820–821. doi:10.1038/nrm3019
Wullschleger S, Loewith R, Hall MN (2006) TOR signaling in growth and metabolism. Cell 124(3):471–484. doi:10.1016/j.cell.2006.01.016
Yamaguchi T, Hempson SJ, Reif GA, Hedge AM, Wallace DP (2006) Calcium restores a normal proliferation phenotype in human polycystic kidney disease epithelial cells. J Am Soc Nephrol JASN 17(1):178–187. doi:10.1681/ASN.2005060645

Chapter 6
Epigenetic Regulation in Cystogenesis

Yu Mi Woo

Abstract Epigenetic regulation refers to heritable changes in gene expression that do not involve any alteration of the DNA sequence. DNA methylation, histone modification, and gene regulation by microRNAs are well-known epigenetic modulations that are closely associated with several cellular processes and diverse disease states, such as cancers, even under precancerous conditions. More recently, several studies have indicated that epigenetic changes may be associated with renal cystic diseases, including autosomal dominant polycystic kidney disease, and the restoration of altered epigenetic factors may become a therapeutic target of renal cystic disease and would be expected to have minimal side effects. This review focuses on recently reported findings on epigenetic and considers the potential of targeting epigenetic regulation as a novel therapeutic approach to control cystogenesis.

Keywords Epigenetic regulation • Cyst • PKD

6.1 Overview of Epigenetic Regulation

Chromatin is a complex formed from DNA, histones, and non-histone proteins in the nucleus. Complicated chromatin remodeling mechanisms keep DNA accessible to transcriptional factors (Felsenfeld and Groudine 2003). Epigenetic regulation refers to the control of gene expression by modulating DNA-protein interactions without any alteration of the genetic code. Epigenetic modulations including DNA methylation, histone modification, and gene regulation by non-coding RNAs including microRNAs (miRNAs) have important roles in several cellular processes and diverse disease states such as cancers even under precancerous conditions (Taby and Issa 2010). Herein, we will provide a brief review of the relatively well understood epigenetic regulation mechanisms: DNA methylation, histone modifications, and miRNAs.

Y.M. Woo (✉)
Molecular Medicine Laboratory, Department of Life systems, Sookmyung Women's University, Cheongpa-ro 47-gil 100, Yongsan-gu, Seoul 04310, South Korea
e-mail: milkmi@sookmyung.ac.kr

J.H. Park, C. Ahn (eds.), *Cystogenesis*, Advances in Experimental Medicine and Biology 933, DOI 10.1007/978-981-10-2041-4_6

6.1.1 Transcriptional Regulation by DNA Methylation

DNA methylation occurs by the covalent addition of a methyl group from S-adenosyl methionine to cytosine residues at cytosine-guanosine dinucleotides (CpGs), which is mediated by DNA methyltransferases (DNMTs) (Bird 2002). During DNA replication, DNMT1, which is known as the maintenance methyltransferase, binds mainly to hemimethylated DNA (Bestor 2000). In contrast, de novo methylation is preferentially mediated by DNMT3a and DNMT3b (Okano et al. 1999). Unmethylated CpGs are grouped in clusters called 'CpG islands' that are located near the promoters of approximately 50 % of all human genes. DNA hypermethylation of CpG islands near the promoter regions silences the expression of specific genes such as tumor suppressor genes or genes coding for miRNAs in cooperation with histone modification (Herman and Baylin 2003).

Although it has been well established that promoter DNA methylation silences gene expression, the effect of gene-body methylation on gene expression is controversial. While the gene-bodies of active genes are preferentially methylated, the methylation level of inactive gene-bodies seems to follow a tissue-specific pattern (Aran et al. 2011). Moreover, several recent studies have shown that the hypermethylation of gene-bodies silenced gene expression, especially that of highly expressed gene groups (Woo et al. 2014; Deaton et al. 2011; Lister et al. 2009).

DNA methylation patterns are intricately associated with patterns of histone modification. Methyl-CpG-binding protein 2 can be recruited to the methylated cytosine (5-methylcytosine) of silenced promoters. Although the exact order of events in gene silencing is still unknown, proteins having methylated DNA-binding domains are linked to large protein complexes containing histone deacetylases (HDACs) and histone methyltransferases (HMTs), which further repress gene transcription by passive demethylation through the administration of DNMT inhibitors such as 5-aza-2′-deoxycytidine (Juttermann et al. 1994).

6.1.2 The Regulatory Mechanism of Histone Modification

Histone modifications as epigenetic marks include histone lysine acetylation, methylation, and phosphorylation and regulate chromatin structure and gene expression (Kouzarides 2007; Zhou et al. 2011). The acetylation of lysine residues on the N-terminal tails of H3 and H4 is mediated by histone acetyltransferases, resulting in the open structure of chromatin. On the other hand, HDACs remove the acetyl groups of lysine residues, leading to a closed chromatin structure (Yoo and Jones 2006). Meanwhile, histone methylation is mediated by HMTs (Sawan and Herceg 2010). There are two families of HMTs, which have specificities for arginine and lysine residues, respectively (Zhang and Reinberg 2001). Histone methylation can be related to either transcriptional activation or repression. Trimethylation of histone H3 at lysine 4 (H3K4me3) is an activating mark while demethylation of histone H3 at lysine 9 (H3K9me2) is an inactivating mark of transcription (Gupta et al. 2010). Histone N-terminal tails are rich in residues with covalent modifications and it has been suggested that these different modifications can affect each other. The arrangement of these marks is termed the histone code (Zhang and Reinberg 2001).

6.1.3 Biogenesis and Mechanism of microRNAs

microRNAs are endogenous small non-coding RNAs of about 22 nucleotides in length and regulate gene expression by targeting the 3′-untranslated region of target genes (Lee et al. 1993). It has been reported that a single miRNA has the potential to regulate multiple mRNA transcripts (Selbach et al. 2008) and one transcript may be regulated by multiple miRNAs (Lindow and Gorodkin 2007). All miRNAs are processed and matured through a multi-step event. At first, miRNA genes are transcribed into long primary transcripts (pri-miRNAs) with one or more stem-loop structures by RNA polymerase II. These pri-miRNAs are cleaved by the nuclear RNase III enzyme, Drosha, into 70–100-nucleotide long precursor miRNAs (pre-miRNAs) that have a hairpin structure with a 3′-overhang. The pre-miRNAs are then transported from the nucleus to the cytosol by exportin-5 and further processed by the cytoplasmic enzyme RNase III Dicer into a mature miRNA duplex. Finally, the mature single-stranded miRNA, which is called the guide strand and is expressed at a much higher level than the other strand, is incorporated into the RNA-induced silencing complex (RISC) (Kim et al. 2009). The other strand, which is called the passenger strand and denoted with an asterisk (*) at the end of its name, is degraded. In some cases, both strands, which are named 5p and 3p, are functional and detected at significant levels (Hu et al. 2009). miRNA-RISC complexes bind to the 3′-untranslated region of the target mRNA by perfect or imperfect base-pairing (Yang and Lai 2011). Animal miRNAs usually have only partial homology with 7 ~ 8 bases of conserved seed sequence of their target genes and may repress translational activity (Grimson et al. 2007).

6.2 Epigenetic Biomarkers in Autosomal Dominant Polycystic Kidney Disease (ADPKD)

6.2.1 DNA Methylation as a Possible Prognostic Biomarker of ADPKD

Aberrant DNA methylation patterns have been closely associated with an abnormal expression of certain genes, ultimately resulting in diverse pathologies such as cancer (Esteller 2005). Intriguingly, because it seems that alterations of DNA methylation are early and tissue-specific events in tumorigenesis (Illingworth et al. 2008), it is possible to apply DNA methylation marks as potential prognostic markers to predict ADPKD progression. Recent studies of genome-wide DNA methylation screening using clinical samples from ADPKD patients have indicated that DNA methylation can be considered as one of the key regulatory mechanisms in the cystogenesis of ADPKD (Woo et al. 2014). In this study, many key regulatory genes involved in cellular transport, calcium signaling, cell morphogenesis, and cell adhesion were markedly hypermethylated, especially in the gene-body region, in ADPKD patients compared with healthy controls. Importantly, hypermethylation of *PKD1* in the gene-body but not in the promoter region in ADPKD kidneys was found to show an inverse correlation with its expression level. In addition, a further recent study identified a novel candidate DNA methylation biomarker of ADPKD. Based on genome-wide DNA methylation profiling, the proximal promoter region of the mucin-like protocadherin (MUPCDH) gene was significantly hypermethylated in ADPKD cystic kidneys compared with healthy control kidneys. *MUPCDH* was specifically downregulated in the cyst-lining epithelial cells of ADPKD. More importantly, a distinctive DNA methylation pattern of the *MUPCDH* promoter was also observed in urine sediments obtained from ADPKD patients with a rapid rate of increase in kidney volume, indicating the possibility that MUPCDH methylation could be used as an epigenetic biomarker of disease progression in ADPKD (Woo et al. 2015). Taken together, these studies suggest that DNA methylation-mediated epigenetic silencing may be one of the mechanisms underlying cyst formation and applied as a biomarker of ADPKD.

6.2.2 Profiling of microRNAs to Develop Biomarkers of ADPKD

The contribution of miRNA to the pathogenesis of ADPKD was investigated in recent studies. Defects in the miRNA-processing enzyme Dicer in maturing renal tubules have been shown to induce tubular and glomerular cysts in Dicer mutant mouse models such as Hoxb7/cre;Dicer$^{f/f}$ and ksp/cre;Dicer$^{f/f}$ (Pastorelli et al. 2009; Patel et al. 2012). Inactivation of the Dicer enzyme leads to abnormal processing of miRNAs including miR-200, which directly targets *Pkd1* (Patel et al. 2012). These studies indicate that miRNAs may play important roles in ADPKD pathogenesis.

In an effort to identify novel miRNA biomarkers and understand the systemic mechanisms underlying ADPKD, miRNA profiling studies have been performed using ADPKD rodent models and ADPKD clinical samples. Pandey et al. performed global gene-expression profiling by microarray in the kidneys of Pkd1-deficient mouse at embryonic days 14.5 and 17.5, and predicted associations of PKD with miR-10a, miR-30a-5p, miR-96, mi-126-5p, miR-182, miR-200a, miR-204, miR-429, and miR-488 using computational methods (Pandey et al. 2011). Dweep et al.

identified 8 upregulated miRNAs (miR-199a-5p, miR-214, miR-146b, miR-21, miR-34a, miR-132, miR-31, and miR-503) in the PKD/Mhm rat model via parallel analysis using mRNA and miRNA microarrays (Dweep et al. 2013).

On the other hand, while most miRNAs are present in the cytosol of cells, many miRNAs are outside the cells and in the body fluids. Many studies have shown that circulating miRNAs are remarkably stable and have distinct expression profiles in different fluids (Zen and Zhang 2012; Weber et al. 2010; Mitchell et al. 2008). Indeed, miRNA profiles and biochemical characteristics were analyzed in urine specimens of ADPKD patients (n = 20), patients with chronic kidney disease of other etiologies (n = 20), and primary cultures of epithelial cells from ADPKD cysts (n = 10), normal adult tubules (n = 8), and fetal tubules (n = 7). In this report, the authors proposed a role for the repression of miR-1 and miR-133b in the pathogenesis of ADPKD and their possible application as biomarkers for monitoring disease progression (Ben-Dov et al. 2014). However, further longitudinal studies of these miRNAs are required for the establishment of their exact roles as potential biomarkers of ADPKD progression.

6.3 Epigenetic Therapy in ADPKD

6.3.1 The Effect on DNMT Inhibitors in ADPKD

In previous studies, hypermethylation of *PKD1* and other genes associated with ion transport and cell adhesion has been identified in ADPKD renal tissue (Woo et al. 2014; Woo et al. 2015). In these reports, the administration of DNMT inhibitors such as 5-aza-2′-deoxycytidine and zebularine significantly retarded cyst growth of MDCK cells embedded in a collagen matrix accompanied by a restoration of gene expression level, indicating that reducing the level of DNA methylation by pharmacological treatment can lead to the re-expression of key gene sets related to cyst formation.

However, targeting DNMT is problematic in that it causes a general reduction of methylation across the genome at random but does not target the reactivation of a specific gene (Gius et al. 2004). Nevertheless, DNA methylation represents an epigenetic mechanism of gene promoter silencing that has been linked to diverse diseases (Costello et al. 2000), and DNMT inhibitors have been shown to attenuate tumorigenicity in a mouse model of neoplasia (Laird et al. 1995).

6.3.2 The Effect on HDAC Inhibitors in ADPKD

To date, evidence from recent studies has demonstrated that histone lysine acetylation may be associated with the pathogenesis of PKD and chronic kidney disease (Van Bodegom et al. 2006; Cao et al. 2009; van Bodegom et al. 2010). Indeed, the administration of HDAC inhibitors retarded cyst growth and preserved renal function in Pkd1 (Zhou et al. 2013; Fan et al. 2012; Li 2011; Liu et al. 2012; Zhou et al. 2015; Zhou et al. 2014) and Pkd2 (Cao et al. 2009) mutant mice.

Trichostatin A, which is a general inhibitor of class I and II HDACs, has been shown to abolish the p53-mediated repression of *Pkd1* gene expression in *Pkd1* mutant mice (Van Bodegom et al. 2006). In addition, HDAC1 induces deactylation of p53, leading to the p53-induced repression of *PKD1* gene transcription (van Bodegom et al. 2010). Intriguingly, the repression of HDAC1 activity through either treatment with HDAC1 inhibitors or Hdac1 siRNA-induced knockdown leads to the retardation of renal cystogenesis in *Pkd2* knockdown zebrafish and *Pkd1* mutant mice (Cao et al. 2009). HDAC5, a class II HDAC, is responsible for polycystin-dependent fluid flow-induced calcium signaling in renal epithelial cells (Xia et al. 2010). The overexpression of HDAC6 has been observed in embryonic renal epithelial cells of *Pkd1* mutant mice (Li 2011). HDAC6 suppresses endocytosis and promotes the degradation of epidermal growth factor receptor, which is a stimulator of cystogenesis through the regulation of α-tubulin acetylation (Deribe et al. 2009). In addition to HDAC6, SIRT2, another member of the class III HDACs family, was upregulated in the inner medullary collecting duct cells of Pkd1 knockdown mice and *Pkd1*-deficient mouse kidney cells. SIRT2 has shown to regulate cilium disassembly via deacetylation of α-tubulin during the normal cell cycle. The SIRT2-HDAC6 complex seems to be involved in α-tubulin deacetylation by binding to α-tubulin (North et al. 2003; Nahhas et al. 2007).

Therefore, these data suggest the potential clinical use of HDAC inhibitors in the treatment of ADPKD. Several HDAC inhibitors have either been approved by the US Food and Drug Administration (FDA) or tested in clinical trials (Cao et al. 2009; Li 2011). In the field of cancer biology, it has been predicted that a combination therapy of HDAC inhibitors and conditional drugs may be more effective than single drug treatments (Yoo and Jones 2006), suggesting that such a combination therapy may also represent an effective strategy for the control of cystogenesis.

6.3.3 microRNAs as Therapeutic Targets in ADPKD

miRNAs have emerged as potential new regulators of ADPKD progression and are considered to be attractive therapeutic targets. Recent studies have demonstrated that miRNA expression patterns may be tightly correlated with the pathogenesis of PKD (Tran et al. 2010; Lee et al. 2008; Pandey et al. 2008; Lakhia et al. 2015; Patel et al. 2013; Patel et al. 2012). miR-15a is down-regulated in cystic kidney tissues and affects cell proliferation and cyst growth by directly targeting Cdc25A, which is known as a cell cycle regulator (Lee et al. 2008). Overexpression of the miR-17 ~ 92 miRNA cluster or miR-21 resulted in the acceleration of cyst growth in PKD mouse models. Interestingly, repression of these miRNAs by the double knockout of either *Kif3α* -miR-17 ~ 92 or *Pkd2*-miR-21 significantly reduced cyst growth compared with *Kif3α*- or *Pkd2*-knockout mice (Patel et al. 2013; Lakhia et al. 2015).

In particular, miR-17 and related miRNA families also represent attractive therapeutic targets, because they repress gene sets that are involved in cyst-promoting pathways such as the mTOR pathway as well as PKD1 (Noureddine et al. 2013).

miRNA-based therapeutic techniques have been suggested, including the inhibition or restoration of the activity of miRNAs using synthetic oligonucleotides called antimiRs and antagomiRs (Patel and Noureddine 2012). The delivery of antagomiRs such as cholesterol-coupled 2′-O-methyl-modified oligoribonucleotides or antisense-locked nucleic acid sequences can be used to inhibit specific miRNA activity. For instance, antimiR-192 was reported to silence miR-192 in several organs including the kidney (Krutzfeldt et al. 2005). Alternatively, in the case of decreased miRNA-induced disease development, miRNA precursors with structures similar to those of endogenous miRNAs can be applied for miRNA-based therapy for ADPKD (Esau and Monia 2007).

Taken together, to identify other potential therapeutic targets among the miRNA profiles associated with ADPKD pathogenesis, careful in vivo studies including consideration of the specificity of delivery and durability of miRNA in the body will be needed in the near future.

References

Aran D, Toperoff G, Rosenberg M, Hellman A (2011) Replication timing-related and gene body-specific methylation of active human genes. Hum Mol Genet 20(4):670–680. doi:10.1093/hmg/ddq513

Ben-Dov IZ, Tan YC, Morozov P, Wilson PD, Rennert H, Blumenfeld JD, Tuschl T (2014) Urine microRNA as potential biomarkers of autosomal dominant polycystic kidney disease progression: description of miRNA profiles at baseline. PLoS One 9(1):e86856. doi:10.1371/journal.pone.0086856

Bestor TH (2000) The DNA methyltransferases of mammals. Hum Mol Genet 9(16):2395–2402

Bird A (2002) DNA methylation patterns and epigenetic memory. Genes Dev 16(1):6–21. doi:10.1101/gad.947102

Cao Y, Semanchik N, Lee SH, Somlo S, Barbano PE, Coifman R, Sun Z (2009) Chemical modifier screen identifies HDAC inhibitors as suppressors of PKD models. Proc Natl Acad Sci U S A 106(51):21819–21824. doi:10.1073/pnas.0911987106

Costello JF, Fruhwald MC, Smiraglia DJ, Rush LJ, Robertson GP, Gao X, Wright FA, Feramisco JD, Peltomaki P, Lang JC, Schuller DE, Yu L, Bloomfield CD, Caligiuri MA, Yates A, Nishikawa R, Su Huang H, Petrelli NJ, Zhang X, O'Dorisio MS, Held WA, Cavenee WK, Plass C (2000) Aberrant CpG-island methylation has non-random and tumour-type-specific patterns. Nat Genet 24(2):132–138. doi:10.1038/72785

Deaton AM, Webb S, Kerr AR, Illingworth RS, Guy J, Andrews R, Bird A (2011) Cell type-specific DNA methylation at intragenic CpG islands in the immune system. Genome Res 21(7):1074–1086. doi:10.1101/gr.118703.110

Deribe YL, Wild P, Chandrashaker A, Curak J, Schmidt MH, Kalaidzidis Y, Milutinovic N, Kratchmarova I, Buerkle L, Fetchko MJ, Schmidt P, Kittanakom S, Brown KR, Jurisica I, Blagoev B, Zerial M, Stagljar I, Dikic I (2009) Regulation of epidermal growth factor receptor trafficking by lysine deacetylase HDAC6. Sci Signal 2(102):ra84. doi:10.1126/scisignal.2000576

Dweep H, Sticht C, Kharkar A, Pandey P, Gretz N (2013) Parallel analysis of mRNA and microRNA microarray profiles to explore functional regulatory patterns in polycystic kidney disease: using PKD/Mhm rat model. PLoS One 8(1):e53780. doi:10.1371/journal.pone.0053780

Esau CC, Monia BP (2007) Therapeutic potential for microRNAs. Adv Drug Deliv Rev 59(2-3):101–114. doi:10.1016/j.addr.2007.03.007

Esteller M (2005) Aberrant DNA methylation as a cancer-inducing mechanism. Annu Rev Pharmacol Toxicol 45:629–656. doi:10.1146/annurev.pharmtox.45.120403.095832

Fan LX, Li X, Magenheimer B, Calvet JP (2012) Inhibition of histone deacetylases targets the transcription regulator Id2 to attenuate cystic epithelial cell proliferation. Kidney Int 81(1):76–85. doi:10.1038/ki.2011.296

Felsenfeld G, Groudine M (2003) Controlling the double helix. Nature 421(6921):448–453. doi:10.1038/nature01411

Gius D, Cui H, Bradbury CM, Cook J, Smart DK, Zhao S, Young L, Brandenburg SA, Hu Y, Bisht KS, Ho AS, Mattson D, Sun L, Munson PJ, Chuang EY, Mitchell JB, Feinberg AP (2004) Distinct effects on gene expression of chemical and genetic manipulation of the cancer epigenome revealed by a multimodality approach. Cancer Cell 6(4):361–371. doi:10.1016/j.ccr.2004.08.029

Grimson A, Farh KK, Johnston WK, Garrett-Engele P, Lim LP, Bartel DP (2007) MicroRNA targeting specificity in mammals: determinants beyond seed pairing. Mol Cell 27(1):91–105. doi:10.1016/j.molcel.2007.06.017

Gupta S, Kim SY, Artis S, Molfese DL, Schumacher A, Sweatt JD, Paylor RE, Lubin FD (2010) Histone methylation regulates memory formation. J Neurosci Off J Soc Neurosci 30(10):3589–3599. doi:10.1523/JNEUROSCI.3732-09.2010

Herman JG, Baylin SB (2003) Gene silencing in cancer in association with promoter hypermethylation. N Engl J Med 349(21):2042–2054. doi:10.1056/NEJMra023075

Hu HY, Yan Z, Xu Y, Hu H, Menzel C, Zhou YH, Chen W, Khaitovich P (2009) Sequence features associated with microRNA strand selection in humans and flies. BMC Genomics 10:413. doi:10.1186/1471-2164-10-413

Illingworth R, Kerr A, Desousa D, Jorgensen H, Ellis P, Stalker J, Jackson D, Clee C, Plumb R, Rogers J, Humphray S, Cox T, Langford C, Bird A (2008) A novel CpG island set identifies tissue-specific methylation at developmental gene loci. PLoS Biol 6(1):e22. doi:10.1371/journal.pbio.0060022

Juttermann R, Li E, Jaenisch R (1994) Toxicity of 5-aza-2′-deoxycytidine to mammalian cells is mediated primarily by covalent trapping of DNA methyltransferase rather than DNA demethylation. Proc Natl Acad Sci U S A 91(25):11797–11801

Kim VN, Han J, Siomi MC (2009) Biogenesis of small RNAs in animals. Nat Rev Mol Cell Biol 10(2):126–139. doi:10.1038/nrm2632

Kouzarides T (2007) Chromatin modifications and their function. Cell 128(4):693–705. doi:10.1016/j.cell.2007.02.005

Krutzfeldt J, Rajewsky N, Braich R, Rajeev KG, Tuschl T, Manoharan M, Stoffel M (2005) Silencing of microRNAs in vivo with 'antagomirs'. Nature 438(7068):685–689. doi:10.1038/nature04303

Laird PW, Jackson-Grusby L, Fazeli A, Dickinson SL, Jung WE, Li E, Weinberg RA, Jaenisch R (1995) Suppression of intestinal neoplasia by DNA hypomethylation. Cell 81(2):197–205

Lakhia R, Hajarnis S, Williams D, Aboudehen K, Yheskel M, Xing C, Hatley ME, Torres VE, Wallace DP, Patel V (2015) MicroRNA-21 aggravates cyst growth in a model of polycystic kidney disease. J Am Soc Nephrol JASN. doi:10.1681/ASN.2015060634

Lee RC, Feinbaum RL, Ambros V (1993) The C. elegans heterochronic gene lin-4 encodes small RNAs with antisense complementarity to lin-14. Cell 75(5):843–854

Lee SO, Masyuk T, Splinter P, Banales JM, Masyuk A, Stroope A, Larusso N (2008) MicroRNA15a modulates expression of the cell-cycle regulator Cdc25A and affects hepatic cystogenesis in a rat model of polycystic kidney disease. J Clin Invest 118(11):3714–3724. doi:10.1172/JCI34922

Li X (2011) Epigenetics and autosomal dominant polycystic kidney disease. Biochim Biophys Acta 1812(10):1213–1218. doi:10.1016/j.bbadis.2010.10.008

Lindow M, Gorodkin J (2007) Principles and limitations of computational microRNA gene and target finding. DNA Cell Biol 26(5):339–351. doi:10.1089/dna.2006.0551

Lister R, Pelizzola M, Dowen RH, Hawkins RD, Hon G, Tonti-Filippini J, Nery JR, Lee L, Ye Z, Ngo QM, Edsall L, Antosiewicz-Bourget J, Stewart R, Ruotti V, Millar AH, Thomson JA, Ren B, Ecker JR (2009) Human DNA methylomes at base resolution show widespread epigenomic differences. Nature 462(7271):315–322. doi:10.1038/nature08514

Liu W, Fan LX, Zhou X, Sweeney WE Jr, Avner ED, Li X (2012) HDAC6 regulates epidermal growth factor receptor (EGFR) endocytic trafficking and degradation in renal epithelial cells. PLoS One 7(11):e49418. doi:10.1371/journal.pone.0049418

Mitchell PS, Parkin RK, Kroh EM, Fritz BR, Wyman SK, Pogosova-Agadjanyan EL, Peterson A, Noteboom J, O'Briant KC, Allen A, Lin DW, Urban N, Drescher CW, Knudsen BS, Stirewalt DL, Gentleman R, Vessella RL, Nelson PS, Martin DB, Tewari M (2008) Circulating microRNAs as stable blood-based markers for cancer detection. Proc Natl Acad Sci U S A 105(30):10513–10518. doi:10.1073/pnas.0804549105

Nahhas F, Dryden SC, Abrams J, Tainsky MA (2007) Mutations in SIRT2 deacetylase which regulate enzymatic activity but not its interaction with HDAC6 and tubulin. Mol Cell Biochem 303(1-2):221–230. doi:10.1007/s11010-007-9478-6

North BJ, Marshall BL, Borra MT, Denu JM, Verdin E (2003) The human Sir2 ortholog, SIRT2, is an NAD+-dependent tubulin deacetylase. Mol Cell 11(2):437–444

Noureddine L, Hajarnis S, Patel V (2013) MicroRNAs and polycystic kidney disease. Drug Discov Today Dis Mod 10(3):e137–e1743. doi:10.1016/j.ddmod.2013.10.001

Okano M, Bell DW, Haber DA, Li E (1999) DNA methyltransferases Dnmt3a and Dnmt3b are essential for de novo methylation and mammalian development. Cell 99(3):247–257

Pandey P, Brors B, Srivastava PK, Bott A, Boehn SN, Groene HJ, Gretz N (2008) Microarray-based approach identifies microRNAs and their target functional patterns in polycystic kidney disease. BMC Genomics 9:624. doi:10.1186/1471-2164-9-624

Pandey P, Qin S, Ho J, Zhou J, Kreidberg JA (2011) Systems biology approach to identify transcriptome reprogramming and candidate microRNA targets during the progression of polycystic kidney disease. BMC Syst Biol 5:56. doi:10.1186/1752-0509-5-56

Pastorelli LM, Wells S, Fray M, Smith A, Hough T, Harfe BD, McManus MT, Smith L, Woolf AS, Cheeseman M, Greenfield A (2009) Genetic analyses reveal a requirement for Dicer1 in the mouse urogenital tract. Mamm Genome Off J Int Mamm Genome Soc 20(3):140–151. doi:10.1007/s00335-008-9169-y

Patel V, Noureddine L (2012) MicroRNAs and fibrosis. Curr Opin Nephrol Hypertens 21(4):410–416. doi:10.1097/MNH.0b013e328354e559

Patel V, Hajarnis S, Williams D, Hunter R, Huynh D, Igarashi P (2012) MicroRNAs regulate renal tubule maturation through modulation of Pkd1. J Am Soc Nephrol JASN 23(12):1941–1948. doi:10.1681/ASN.2012030321

Patel V, Williams D, Hajarnis S, Hunter R, Pontoglio M, Somlo S, Igarashi P (2013) miR-17~92 miRNA cluster promotes kidney cyst growth in polycystic kidney disease. Proc Natl Acad Sci U S A 110(26):10765–10770. doi:10.1073/pnas.1301693110

Sawan C, Herceg Z (2010) Histone modifications and cancer. Adv Genet 70:57–85. doi:10.1016/B978-0-12-380866-0.60003-4

Selbach M, Schwanhausser B, Thierfelder N, Fang Z, Khanin R, Rajewsky N (2008) Widespread changes in protein synthesis induced by microRNAs. Nature 455(7209):58–63. doi:10.1038/nature07228

Taby R, Issa JP (2010) Cancer epigenetics. CA Cancer J Clin 60(6):376–392. doi:10.3322/caac.20085

Tran U, Zakin L, Schweickert A, Agrawal R, Doger R, Blum M, De Robertis EM, Wessely O (2010) The RNA-binding protein bicaudal C regulates polycystin 2 in the kidney by antagonizing miR-17 activity. Development 137(7):1107–1116. doi:10.1242/dev.046045

Van Bodegom D, Saifudeen Z, Dipp S, Puri S, Magenheimer BS, Calvet JP, El-Dahr SS (2006) The polycystic kidney disease-1 gene is a target for p53-mediated transcriptional repression. J Biol Chem 281(42):31234–31244. doi:10.1074/jbc.M606510200

van Bodegom D, Roessingh W, Pridjian A, El Dahr SS (2010) Mechanisms of p53-mediated repression of the human polycystic kidney disease-1 promoter. Biochim Biophys Acta 1799(7):502–509. doi:10.1016/j.bbagrm.2010.04.001

Weber JA, Baxter DH, Zhang S, Huang DY, Huang KH, Lee MJ, Galas DJ, Wang K (2010) The microRNA spectrum in 12 body fluids. Clin Chem 56(11):1733–1741. doi:10.1373/clinchem.2010.147405

Woo YM, Bae JB, Oh YH, Lee YG, Lee MJ, Park EY, Choi JK, Lee S, Shin Y, Lyu J, Jung HY, Lee YS, Hwang YH, Kim YJ, Park JH (2014) Genome-wide methylation profiling of ADPKD identified epigenetically regulated genes associated with renal cyst development. Hum Genet 133(3):281–297. doi:10.1007/s00439-013-1378-0

Woo YM, Shin Y, Hwang JA, Hwang YH, Lee S, Park EY, Kong HK, Park HC, Lee YS, Park JH (2015) Epigenetic silencing of the MUPCDH gene as a possible prognostic biomarker for cyst growth in ADPKD. Sci Rep 5:15238. doi:10.1038/srep15238

Xia S, Li X, Johnson T, Seidel C, Wallace DP, Li R (2010) Polycystin-dependent fluid flow sensing targets histone deacetylase 5 to prevent the development of renal cysts. Development 137(7):1075–1084. doi:10.1242/dev.049437

Yang JS, Lai EC (2011) Alternative miRNA biogenesis pathways and the interpretation of core miRNA pathway mutants. Mol Cell 43(6):892–903. doi:10.1016/j.molcel.2011.07.024

Yoo CB, Jones PA (2006) Epigenetic therapy of cancer: past, present and future. Nat Rev Drug Discov 5(1):37–50. doi:10.1038/nrd1930

Zen K, Zhang CY (2012) Circulating microRNAs: a novel class of biomarkers to diagnose and monitor human cancers. Med Res Rev 32(2):326–348. doi:10.1002/med.20215

Zhang Y, Reinberg D (2001) Transcription regulation by histone methylation: interplay between different covalent modifications of the core histone tails. Genes Dev 15(18):2343–2360. doi:10.1101/gad.927301

Zhou VW, Goren A, Bernstein BE (2011) Charting histone modifications and the functional organization of mammalian genomes. Nat Rev Genet 12(1):7–18. doi:10.1038/nrg2905

Zhou X, Fan LX, Sweeney WE Jr, Denu JM, Avner ED, Li X (2013) Sirtuin 1 inhibition delays cyst formation in autosomal-dominant polycystic kidney disease. J Clin Invest 123(7):3084–3098. doi:10.1172/JCI64401

Zhou X, Fan LX, Li K, Ramchandran R, Calvet JP, Li X (2014) SIRT2 regulates ciliogenesis and contributes to abnormal centrosome amplification caused by loss of polycystin-1. Hum Mol Genet 23(6):1644–1655. doi:10.1093/hmg/ddt556

Zhou X, Fan LX, Peters DJ, Trudel M, Bradner JE, Li X (2015) Therapeutic targeting of BET bromodomain protein, Brd4, delays cyst growth in ADPKD. Hum Mol Genet 24(14):3982–3993. doi:10.1093/hmg/ddv136

Part III
Therapeutic Approaches and Diagnostic Markers for ADPKD

Chapter 7
Validation of Effective Therapeutic Targets for ADPKD Using Animal Models

Yu Mi Woo, Je Yeong Ko, and Eun Ji Lee

Abstract Various polycystic kidney disease (PKD) animal models including *Pkd1*- or *Pkd2*-deficient mice have been developed and efficiently utilized to identify novel therapeutic targets as well as elucidate multiple mechanisms of cyst formation in PKD. Based on several successful in vivo studies, preclinical approaches using PKD animal models would shed light on the development of potential therapeutic strategies for PKD. Here, we provide an update on the current evidence obtained by the in vivo evaluation of PKD therapeutic candidates and discuss the effect of therapeutic targets.

Keywords PKD • Animal model • Therapeutic target

7.1 Various Polycystic Kidney Disease (PKD) Animal Models Are Available to Reveal the Biological Functions of PKD-Causing Genes

The initial development of PKD is driven by an increase of cell proliferation. However, depending on the disease progression, dysregulated apoptosis, differentiation, fibrosis, and inflammation can also occur, indicating that PKD is a complex disease induced by defects of multiple signaling pathways. Based on this characteristic of PKD, many research groups have developed PKD mice models to understand the physiological mechanisms of PKD development and screen effective therapeutic targets for curing PKD. Rodent models of PKD share common pathogenic phenotypes, including cyst formation in multiple nephron segments and an increase of cell proliferation, but display different characteristics in the progression

Y.M. Woo (✉) • J.Y. Ko • E.J. Lee
Molecular Medicine Laboratory, Department of Life systems, Sookmyung Women's University, Cheongpa-ro 47-gil 100, Yongsan-gu, Seoul 04310, South Korea
e-mail: milkmi@sm.ac.kr; jeyeong@sm.ac.kr; eunji8902@sm.ac.kr

J.H. Park, C. Ahn (eds.), *Cystogenesis*, Advances in Experimental Medicine and Biology 933, DOI 10.1007/978-981-10-2041-4_7

of cyst formation, life span, and renal cilia phenotypes. In this section, the morphological features and signaling alterations of well-established PKD rodent models are introduced.

7.1.1 *Pkd1 or Pkd2-Targeted Mouse Models*

Mutation of the *PKD1* gene is known as a representative cause of the development of human PKD and the most commonly inherited mutation of *PKD* (Kim et al. 2009). Therefore, *Pkd1*-targeted mice were produced to evaluate the biological function of Pkd1 in vivo. While *Pkd1* constitutive knockout mice show embryonic lethality accompanied by kidney cysts, liver cysts, and abnormal cardiovascular and skeletal development (Boulter et al. 2001), mouse models of the kidney-specific inactivation of *Pkd1* usually survive until birth (Shibazaki et al. 2008). Kidneys conditionally targeted by the *Pkd1* gene show rapid cyst formation from postnatal day 1 (P1) to P14 with an increase of cell proliferation followed by activation of the MAPK/ERK pathway (Shibazaki et al. 2008). Another gene mutated in human PKD is *PKD2*, which causes approximately 15 % of familial autosomal dominant PKD (ADPKD) cases (Kim et al. 2009). To reveal the physiological effect of *Pkd2* inactivation, various PKD mice targeted by *Pkd2* have been generated. The *Pkd2* homozygous knockout mutant mice exhibit embryonic lethality like mice homozygous for *Pkd1* and show body edema, cardiac defects, and cysts of the kidney and pancreas (Kim et al. 2009; Wu et al. 2000). In addition to *Pkd2* constitutive knockout mice, *PKD2* transgenic mice were generated (Park et al. 2009). Histological analysis of these transgenic mice showed that renal cysts originated from a range of nephron segments at 18 months of age (Park et al. 2009). Also, the activation of B-Raf/Mek/Erk signaling was observed in the cystic kidneys of this transgenic mouse model (Park et al. 2009). These polycystin-targeted mice models show that the polycystin proteins play a role in organogenesis during embryonic development and that defects of polycystin induce the cystic kidney phenotype via activation of the MAPK/ERK pathway, leading to an increase of cell proliferation.

7.1.2 *PKD Mouse Models Targeted by IFT-Related Genes*

The first PKD mouse model showing a relationship between ciliary dysfunction and PKD development was the oak ridge polycystic kidney (ORPK) mouse induced by mutation of *Ift88* (Tg737, Polaris), which belongs to the IFT-B complex (Lehman et al. 2008). This model shows a number of abnormal phenotypes induced by ciliary malfunction. It has been reported that renal cysts, hydrocephalus, pancreatic abnormalities, and aberrant patterning of skeletal structure are observed in this model (Cano et al. 2004; Banizs et al. 2005; Ko and Park 2013). Furthermore, a reduction in the number of ciliated cells and abnormal ciliary structure were observed in the pancreatic and renal cells of ORPK mice (Cano et al. 2004; Pazour et al. 2000).

In addition to the ORPK mouse model, various PKD mouse models induced by the inactivation of IFT-related genes have been developed. One of the PKD mouse models targeting the IFT complex B subunit was induced by the specific inactivation of *Ift20* in renal collecting duct cells (Jonassen et al. 2008). This model shows severe and rapid renal cyst progression with a complete loss of cilia, leading to the alteration of Wnt signaling (Jonassen et al. 2008). In addition to the Ift20-targeted mouse model, Ift25 and Ift27, which belong to the IFT-B complex, are constitutively inactivated in vivo. Interestingly, although Ift25 and Ift27 are subunits of IFT complex B, which is involved in cilia assembly, the inactivation of these two genes was found to have no effect on cilia assembly (Keady et al. 2012; Eguether et al. 2014). In these models, the phenotype of cilia appears normal, but they display multiple developmental defects such as skeletal malfunctions, omphaloceles, and polydactyly as well as an alteration of Hh signaling (Keady et al. 2012; Eguether et al. 2014). Not only IFT complex B, but also a IFT complex A-targeted mouse have been generated. A representative PKD animal model induced by the inactivation of IFT complex A is a mouse with a conditional allele for *Ift140* in the renal collecting duct cells (Jonassen et al. 2012). In general, subunits that belong to IFT complex A are involved in cilia disassembly, so it is conceivable that the inactivation of *Ift140* might induce an increase of cilia length. However, severe shortening or absence of primary cilia was observed in the *Ift140*-deleted renal collecting duct cells with a PKD phenotype (Jonassen et al. 2012). These findings suggest that a normal ciliary function is important for the maintenance of homeostasis in renal epithelial cells and that defects of ciliary structure or function contribute to the development of PKD through an increase of cell proliferation.

7.1.3 Juvenile Cystic Kidney & Congenital Polycystic Kidney Mice

Juvenile cystic kidney (jck) mice are produced by a missense mutation of the *Nek8* gene (Liu et al. 2002). This mouse model shows renal cysts in multiple nephron segments and a life span of approximately 20–25 weeks (Nagao et al. 2012). The protein product of this mutated gene is observed in the entire length of the primary cilia in kidney, and it results in the abnormal localization of polycystins in lengthened primary cilia in the kidneys of jck mice (Sohara et al. 2008; Smith et al. 2006). Interestingly, the kidney phenotype of jck mice displays gender dimorphism in the progression of cyst formation, with a more severe phenotype in male mice because of gonadal hormones (Smith et al. 2006).

The congenital polycystic kidney (cpk) mouse is one of the PKD models with a mutation of the *Cys1* gene encoding cystin protein, which is localized to the primary cilia (Hou et al. 2002). Most cysts observed in cpk mice are derived from the collecting ducts and proximal tubules and are accompanied by an increased expression of proto-oncogenes and growth factors together with an alteration in the expression of genes associated with cell adhesion (Hou et al. 2002; Ko and Park 2013; Rocco et al. 1992; Nakamura et al. 1993).

7.1.4 Han:SPRD Cy Rat Model

The Han:SPRD Cy rat model is caused by a missense mutation of the *Pkdr1* (also called Cy and Anks6) gene (Nagao et al. 2010). The protein SamCystin that is encoded by the Pkdr1 gene is mainly expressed in the early postnatal kidney and proximal tubules (Nagao et al. 2010). A point mutation of the Pkdr1 gene results in the aberrant expression and mislocalization of SamCystin in this rat model (Nagao et al. 2010). Kidneys of heterozygous mutant rats (Cy/+) show a mild progression of the PKD phenotype compared with homozygous mutant rats (Cy/Cy) (Nagao et al. 2003). In addition, the Han:SPRD Cy rat model displays a gender-specific kidney phenotype. The kidneys of male Cy/+ rats display a more severe renal cystic phenotype compared with that of the female rats, which affects to average life span of both the males and females of the Cy/+ rat model (Nagao et al. 2003).

7.2 Potential Candidate Targets for PKD Treatment

At present, there are no FDA-approved therapies for the treatment of ADPKD. Nevertheless, recent studies have suggested a number of promising targets and molecular pathways related to cystogenesis, providing new insights into potential therapeutic interventions. The main treatment approaches attempted in ADPKD have focused on inhibiting cystic cell proliferation and fluid secretion (Bukanov et al. 2012; Calvet 2008; Yang et al. 2008; Chang and Ong 2012). More currently, inhibition of the renin-angiotensin-aldosterone system, targeting ciliary function, membrane glycosphingolipids, extracellular matrix, and epigenetic restoration, has also been under investigation (Natoli et al. 2010; Elliott et al. 2011; Li 2011). Here, we present a review of candidate ADPKD drugs and current trials according to the drug targets in PKD rodent models, as follows.

7.2.1 Cyclic AMP (cAMP)-Dependent Signaling Inhibitors

cAMP is a well-known regulator involved in cyst fluid accumulation (Wallace et al. 2001; Sullivan et al. 1998), and an elevated level of cAMP stimulates the activation of the B-Raf/MEK/ERK pathway in ADPKD (Yamaguchi et al. 2003). It has been reported that a number of agonists targeting the vasopressin and somatostatin pathways can result in cAMP accumulation (Gattone et al. 2003; Masyuk et al. 2007).

7.2.1.1 Vasopressin V2 Receptor Antagonist

The vasopressin receptor (V2R) on collecting ducts binds to vasopressin and increases cAMP accumulation by activating adenylyl cyclase. The vasopressin V2R antagonists, OPC-31260 and OPC-41061 (tolvaptan), have been shown to reduce renal cAMP and cystogenesis in four rodent models of renal cystic disease (cpk mice, pcy mice, PCK rats, and Pkd2ws25/- mice) (Torres et al. 2004; Wang et al. 2008; Gattone et al. 2003; Gattone et al. 1999). Tolvaptan is effective in the treatment of hypervolemic or euvolemic hyponatremia and congestive heart failure (Irazabal et al. 2011). These promising preclinical results have translated into clinical trials under the Tolvaptan Efficacy and Safety in Management of PKD and Outcomes (TEMPO) 3:4 program (Torres et al. 2011; Torres et al. 2012). The TEMPO 3:4 trial was designed as a 3-year multicenter randomized placebo-controlled trial (n = 1445) investigating the progression of changes in total kidney volume (TKV), PKD complications and drug safety. In ADPKD patients who were treated with tolvaptan for 3 years, the rate of TKV increase was reduced by almost 50 % compared with that in the placebo group (2.8 % vs. 5.5 % per year, $p < 0.001$). Tolvaptan was also shown to ameliorate the decline of renal function (Torres et al. 2012). This result is consistent with the findings of a recent study that evaluated the efficacy of tolvaptan in the Japanese sub-population (n = 177) (Muto et al. 2015). However, the long-term administration of tolvaptan caused reduced tolerability and significant adverse effects. For example, after the discontinuation of tolvaptan, 8.3 % of patients in the treatment group had severe aquaresis and an elevation of aminotransferase enzyme concentrations, indicating the potential for acute liver failure, and TKV progression continued at the same rate as before therapy (Torres et al. 2012). Overall, tolvaptan is the first pharmacotherapeutic intervention to have been demonstrated to have a therapeutic benefit in ADPKD (Baur and Meaney 2014). At present, the TEMPO 4:4 trial is underway in the USA and tolvaptan has been approved in Europe and Japan for the pharmacological treatment of ADPKD.

7.2.1.2 Somatostatin Analogs

The somatostatin agonist octreotide was shown to be effective in slowing the progression of liver and kidney cystic disease in a small group of ADPKD patients (Ruggenenti et al. 2005) and the PCK rat model (Masyuk et al. 2007). The activation of somatostatin SSTR2 receptor, which is expressed in the kidneys, by octreotide significantly decreased the intracellular level of cAMP, consequently slowing cyst growth and disease progression (Hogan et al. 2010). Ruggenenti et al. demonstrated that octreotide decreased TKV in 12 patients with ADPKD in Italy (Ruggenenti et al. 2005). Although it was a short-term pilot study in a small group, they observed a reduction in the rate of TKV increase and cyst size with only mild adverse events such as gastrointestinal disorder. In a follow-up paper, the beneficial effect of octreotide was evaluated in a long-term, randomized, placebo-controlled and multicenter trial (Caroli et al. 2013). In this study, adult ADPKD patients (with an estimated glomerular filtration rate > 40 mL/min per 1.73 m^2) were randomly divided into two groups and treated with two 20 mg intramuscular injections of octreotide longacting release (LAR) (n = 40) or 0.9 % (v/v) sodium chloride solution (n = 39) every 28 days for 3 years. As a result, at the 1-year follow-up, mean TKV increased significantly less in the octreotide-LAR group than it did in the placebo group (46.2 mL vs. 143.7 mL, p = 0.032). However, at the 3-year follow-up, the mean TKV was shown to be significantly different between the two treatment groups (220.1 mL vs. 454.3 mL, p = 0.25). This result indicates the probable occurrence of tachyphylaxis due to the downregulation or desensitization of somatostatin receptors (Hogan et al. 2012; Caroli et al. 2013). Notably, the initial short-term reduction in glomerular filtration rate (GFR) showed a correlation with the subsequent decline of GFR in the octreotide-LAR group, suggesting that the participants who had larger initial reductions in GFR appeared to show a slower long-term progression towards renal failure while being treated with somatostatin (Caroli et al. 2013). Overall, somatostatin analogues were shown to be relatively safe and well tolerated in all participants compared with previous ADPKD trials. At present, a follow-up study, the Developing Interventions to Halt Progression of ADPKD 1 (DIPAK1) Study, which was designed to examine the efficacy of another somatostatin analogue, lanreotide, on renal function in ADPKD, has been conducted mostly in Europe (Meijer et al. 2014).

7.2.2 Mammalian Target of Rapamycin (mTOR) Inhibitors

mTOR is a serine/threonine kinase that is involved in the promotion of cell proliferation and cell division as well as transcription and protein synthesis. Intriguingly, the mTOR signaling pathway is abnormally upregulated in the cyst-lining epithelial cells of ADPKD mouse models, possibly due to a loss of regulation by PC1 (Wahl et al. 2006; Wander et al. 2011). Rapamycin, also known as sirolimus, inhibits mTOR's kinase activity by binding to FK506-binding protein (Sabers et al. 1995). In preclinical studies, mTOR inhibitors including sirolimus and everolimus were shown to be highly effective in decreasing renal cystogenesis and improving kidney function in several rodent models of ADPKD (Shillingford et al. 2006; 2010; Wahl et al. 2006). However, two key randomized, Phase II trials of the studies evaluating mTOR pathway inhibition have failed to demonstrate the therapeutic efficacy of either drug on either TKV or estimated glomerular filtration rate (Torres et al. 2010; Serra et al. 2010; Walz et al. 2010). In addition, both studies showed that treatment with mTOR inhibitors led to therapy-specific side effects including immunosuppression, diarrhea, acne, and mucositis as well as being limited by an inadequate degree of mTOR inhibition (Watnick and Germino 2010). Therefore, it should be considered that the doses of these drugs that were shown to significantly reduce cyst growth in several rodent models were high (approximately 10-fold higher than the doses used in clinical trials) (Novalic et al. 2012). Efforts to overcome the systemic toxicity of mTOR inhibitors are also being made to enhance the drug specificity to the kidney. One possible approach is using folate-conjugated drugs as candidates for kidney-specific targeting, because folate receptors are overexpressed in the apical membranes of proximal tubule cells. In fact, a recent study demonstrated that treatment with folate-conjugated rapamycin (0.3 mol/kg per day) effectively reduced renal cyst development and preserved renal function without adverse events in the bpk mouse model (Shillingford et al. 2012).

7.2.3 Statins (HMG CoA Reductase Inhibitors)

Statins are widely used to reduce cholesterol in clinical settings by inhibiting the enzyme HMG-CoA reductase. Statins were also shown to decrease renal cystogenesis and improve renal function in the Han:SPRD rat model (Gile et al. 1995; Zafar et al. 2007). Recently, it has been reported that 110 young adults with ADPKD were randomly assigned to treatment with pravastatin or placebo for 3 years to determine the effect of pravastatin in ADPKD. Significant effects on the primary outcomes were shown with a significant decrease in the rate of TKV increase over the study period (pravastatin: 23 % vs. placebo: 31 %, $p=0.02$) (Cadnapaphornchai et al. 2014). However, it is difficult to determine the efficacy of pravastatin because there were no significant changes of renal function or urinary protein excretion between the pravastatin and placebo treated groups in a randomized clinical trial of 49 adults with ADPKD for 2 years (Fassett et al. 2010).

7.3 Other Pre-Clinical Trials That Have Attempted to Identify the Potential Therapeutic Targets of ADPKD

Other therapeutic strategies targeting cell proliferation have been investigated, including direct inhibition of the cell proliferation-regulating proteins that are involved in the Raf/MEK/ERK signaling pathway. Sorafenib, a non-selective Raf inhibitor that finally reduces ERK activation, completely inhibited in vitro cyst growth in human ADPKD cystic cells cultured within a three dimensional collagen gel (Yamaguchi et al. 2010). However, unexpectedly, the administration of sorafenib to *Pkd2* conditional knockout mice promotes liver cyst growth (Spirli et al. 2012). Another group reported that a different small molecule Raf inhibitor (PXL5568) retarded cyst expansion without an improvement in renal function in the Han:SPRD rat model (Buchholz et al. 2011). In addition, PD184352 of MEK inhibitors reduced cyst development and disease progression in the pcy mouse model, but UO126, which is another MEK inhibitor (Omori et al. 2006), was not shown to significantly alter cyst growth in the Pkd1 conditional knockout mouse model (Shibazaki et al. 2008). For another trial, metformin, an AMP-activated protein kinase (AMPK) activator, repressed cyst growth of MDCK cells cultured in vitro in collagen gels and in vivo *Pkd1* conditional knockout mice by activating AMPK and suppressing mTOR and CFTR (Takiar et al. 2011). Furthermore, alterations of glycosphingolipid metabolism with increased GlcCer may have an important role in promoting cyst development. Inhibition of the synthesis of GlcCer blocked cell cycle progression and proliferation by repressing the Akt/mTOR pathway in ADPKD mouse models (Natoli et al. 2010).

Renal cysts are mainly caused by the dysregulation of cell proliferation followed by imbalanced calcium influx as well as the malfunction of the PC1-PC2 protein complex. A large majority of pre-clinical trials have been focused on cAMP signaling, vasopressin-V2R, and the signaling pathways centered on the mTOR or MAP kinases. However, accumulating evidence has suggested that ADPKD progression seems to be influenced by the accumulated factors inside the cysts released by cyst-lining epithelia (Ye et al. 1992; Gardner et al. 1991). Fluids secretion is mainly accelerated by abnormal chloride efflux into the cyst cavity via cystic fibrosis transmembrane conductance regulator (CFTR) or other specific transporters (Miranda et al. 2013). Approaches targeting CFTR or its regulating mechanisms using natural compounds have been attempted in several pre-clinical trials (Yuajit and Chatsudthipong 2016). Among them, steviol, a natural compound firstly isolated from the plant Stevia rebaudiana, effectively slowed down cyst development in ADPKD mice. The rodent model used was *Pkd1*$^{f/f}$: *Pkhd1-cre*, in which *Pkd1* is conditionally knocked-out only in kidney tubular cells, leading to ADPKD. Treatment with steviol delayed the growth of renal cysts in *Pkd1*$^{f/f}$: *Pkhd1-cre* with enhanced renal function as indicated by reduced blood urea nitrogen and creatine levels. The specific mechanisms by which steviol inhibits disease progression were found to be mediated by the CFTR signaling pathway followed by the activation of AMPK. A lowered expression of CFTR subsequently inhibited fluids secretion as well as cell proliferation in steviol-injected mice, and it finally attenuated the disease phenotypes (Yuajit et al. 2014). Some other trials have targeted inflammation or fibrosis, which commonly accompany the progression of ADPKD. Among the tested interventions, Angiotensin-converting enzyme inhibitors have effectively ameliorated the renal cysts development and improved renal function in Han:SPRD rats (Keith et al. 1994; Zafar et al. 2007). Angiotensin essentially stimulates the production of pro-inflammatory factors as well as cell proliferation; therefore, inhibition of its synthesis reasonably led to an alleviation of the disease. The other drug that was evaluated for targeting inflammation in ADPKD was pyrrolidine dithiocarbamate, which has both anti-inflammatory and anti-proliferative effects. Treatment with this agent clearly resulted in a decreased TKV of male Lewis polycystic kidney rats, thereby reducing the ratio of kidney weight to total body weight by about 25 %. However, no changes in cell proliferation, interstitial inflammation, and fibrosis occurred, leading to no effects on renal function (Ta et al. 2014). Apoptosis-regulating mechanisms have been also suggested as another potential target.

Inhibition of caspase-3 activity via treatment with IDN-8050 for 5 weeks resulted in reduced kidney enlargement as well as cysts volume density by 44 % and 29 %, respectively, in Han: SPRD rat model. It led to enhanced renal function with down-regulation both of cell proliferation and apoptosis. Those therapeutic effects have been observed with only three-hour treatment with the same drug as well, which means that long-term administration is not necessarily required for effective alleviation of the disease phenotype (Tao et al. 2005a; b). The indirect effect of targeting apoptosis to inhibit PKD progression has been also been tested by injection of CDK inhibitor roscovitine into cpk and jck mice. Mice treated with roscovitine showed delayed renal cyst development with blockade of cell cycle as well as apoptosis (Bukanov et al. 2006). Besides, a pre-clinical study using lovastin has revealed its therapeutic effects on the metabolic distributions in Han: SPRD rats. Lovastin is a lipid-lowering therapeutic medication and its use in the treatment of the Han:SPRD rat model led to the alleviation of renal cysts with enhanced renal function as well as metabolic alterations (Klawitter et al. 2013). Finally, dietary modulation has been recently suggested as another novel potential therapeutic option for ADPKD. Interestingly, food restriction effectively delayed ADPKD progression with a reduction in the volume of renal cysts, interstitial inflammation, and fibrosis. These changes were mediated by regulating mTOR and AMPK activities (Warner et al. 2015). In these regards, targeting additional disease-stimulating factors other than the main mechanisms that initiate the disease could be another strategy to identify novel therapeutic targets for ADPKD.

References

Banizs B, Pike MM, Millican CL, Ferguson WB, Komlosi P, Sheetz J, Bell PD, Schwiebert EM, Yoder BK (2005) Dysfunctional cilia lead to altered ependyma and choroid plexus function, and result in the formation of hydrocephalus. Development 132(23):5329–5339. doi:10.1242/dev.02153

Baur BP, Meaney CJ (2014) Review of tolvaptan for autosomal dominant polycystic kidney disease. Pharmacotherapy 34(6):605–616. doi:10.1002/phar.1421

Boulter C, Mulroy S, Webb S, Fleming S, Brindle K, Sandford R (2001) Cardiovascular, skeletal, and renal defects in mice with a targeted disruption of the Pkd1 gene. Proc Natl Acad Sci U S A 98(21):12174–12179. doi:10.1073/pnas.211191098

Buchholz B, Klanke B, Schley G, Bollag G, Tsai J, Kroening S, Yoshihara D, Wallace DP, Kraenzlin B, Gretz N, Hirth P, Eckardt KU, Bernhardt WM (2011) The Raf kinase inhibitor PLX5568 slows cyst proliferation in rat polycystic kidney disease but promotes renal and hepatic fibrosis. Nephrol Dial Transplant Off Publ Eur Dial Transplant Assoc Eur Renal Assoc 26(11):3458–3465. doi:10.1093/ndt/gfr432

Bukanov NO, Smith LA, Klinger KW, Ledbetter SR, Ibraghimov-Beskrovnaya O (2006) Long-lasting arrest of murine polycystic kidney disease with CDK inhibitor roscovitine. Nature 444(7121):949–952. doi:10.1038/nature05348

Bukanov NO, Moreno SE, Natoli TA, Rogers KA, Smith LA, Ledbetter SR, Oumata N, Galons H, Meijer L, Ibraghimov-Beskrovnaya O (2012) CDK inhibitors R-roscovitine and S-CR8 effectively block renal and hepatic cystogenesis in an orthologous model of ADPKD. Cell Cycle 11(21):4040–4046. doi:10.4161/cc.22375

Cadnapaphornchai MA, George DM, McFann K, Wang W, Gitomer B, Strain JD, Schrier RW (2014) Effect of pravastatin on total kidney volume, left ventricular mass index, and microalbuminuria in pediatric autosomal dominant polycystic kidney disease. Clin J Am Soc Nephrol CJASN 9(5):889–896. doi:10.2215/CJN.08350813

Calvet JP (2008) Strategies to inhibit cyst formation in ADPKD. Clin J Am Soc Nephrol CJASN 3(4):1205–1211. doi:10.2215/CJN.05651207

Cano DA, Murcia NS, Pazour GJ, Hebrok M (2004) Orpk mouse model of polycystic kidney disease reveals essential role of primary cilia in pancreatic tissue organization. Development 131(14):3457–3467. doi:10.1242/dev.01189

Caroli A, Perico N, Perna A, Antiga L, Brambilla P, Pisani A, Visciano B, Imbriaco M, Messa P, Cerutti R, Dugo M, Cancian L, Buongiorno E, De Pascalis A, Gaspari F, Carrara F, Rubis N, Prandini S, Remuzzi A, Remuzzi G, Ruggenenti P, ALADIN study group (2013) Effect of longacting somatostatin analogue on kidney and cyst growth in autosomal dominant polycystic kidney disease (ALADIN): a randomised, placebo-controlled, multicentre trial. Lancet 382(9903):1485–1495. doi:10.1016/S0140-6736(13)61407-5

Chang MY, Ong AC (2012) Mechanism-based therapeutics for autosomal dominant polycystic kidney disease: recent progress and future prospects. Nephron Clin Pract 120(1):c25–c34; discussion c35. doi:10.1159/000334166

Eguether T, San Agustin JT, Keady BT, Jonassen JA, Liang Y, Francis R, Tobita K, Johnson CA, Abdelhamed ZA, Lo CW, Pazour GJ (2014) IFT27 links the BBSome to IFT for maintenance of the ciliary signaling compartment. Dev Cell 31(3):279–290. doi:10.1016/j.devcel.2014.09.011

Elliott J, Zheleznova NN, Wilson PD (2011) c-Src inactivation reduces renal epithelial cell-matrix adhesion, proliferation, and cyst formation. Am J Physiol Cell Physiol 301(2):C522–C529. doi:10.1152/ajpcell.00163.2010

Fassett RG, Coombes JS, Packham D, Fairley KF, Kincaid-Smith P (2010) Effect of pravastatin on kidney function and urinary protein excretion in autosomal dominant polycystic kidney disease. Scand J Urol Nephrol 44(1):56–61. doi:10.3109/00365590903359908

Gardner KD Jr, Burnside JS, Elzinga LW, Locksley RM (1991) Cytokines in fluids from polycystic kidneys. Kidney Int 39(4):718–724

Gattone VH 2nd, Maser RL, Tian C, Rosenberg JM, Branden MG (1999) Developmental expression of urine concentration-associated genes and their altered expression in murine infantile-type polycystic kidney disease. Dev Genet 24(3–4):309–318. doi:10.1002/(SICI)1520-6408(1999)24:3/4<309::AID-DVG14>3.0.CO;2-5

Gattone VH 2nd, Wang X, Harris PC, Torres VE (2003) Inhibition of renal cystic disease development and progression by a vasopressin V2 receptor antagonist. Nat Med 9(10):1323–1326. doi:10.1038/nm935

Gile RD, Cowley BD Jr, Gattone VH 2nd, O'Donnell MP, Swan SK, Grantham JJ (1995) Effect of lovastatin on the development of polycystic kidney disease in the Han:SPRD rat. Am J Kidney Dis Off J Nat Kidney Found 26(3):501–507

Hogan MC, Masyuk TV, Page LJ, Kubly VJ, Bergstralh EJ, Li X, Kim B, King BF, Glockner J, Holmes DR 3rd, Rossetti S, Harris PC, LaRusso NF, Torres VE (2010) Randomized clinical trial of long-acting somatostatin for autosomal dominant polycystic kidney and liver disease. J Am Soc Nephrol JASN 21(6):1052–1061. doi:10.1681/ASN.2009121291

Hogan MC, Masyuk TV, Page L, Holmes DR 3rd, Li X, Bergstralh EJ, Irazabal MV, Kim B, King BF, Glockner JF, Larusso NF, Torres VE (2012) Somatostatin analog therapy for severe polycystic liver disease: results after 2 years. Nephrol Dial Transplant Off Publ Eur Dial Transplant Assoc Eur Renal Assoc 27(9):3532–3539. doi:10.1093/ndt/gfs152

Hou X, Mrug M, Yoder BK, Lefkowitz EJ, Kremmidiotis G, D'Eustachio P, Beier DR, Guay-Woodford LM (2002) Cystin, a novel cilia-associated protein, is disrupted in the cpk mouse model of polycystic kidney disease. J Clin Invest 109(4):533–540. doi:10.1172/JCI14099

Irazabal MV, Torres VE, Hogan MC, Glockner J, King BF, Ofstie TG, Krasa HB, Ouyang J, Czerwiec FS (2011) Short-term effects of tolvaptan on renal function and volume in patients with autosomal dominant polycystic kidney disease. Kidney Int 80(3):295–301. doi:10.1038/ki.2011.119

Jonassen JA, San Agustin J, Follit JA, Pazour GJ (2008) Deletion of IFT20 in the mouse kidney causes misorientation of the mitotic spindle and cystic kidney disease. J Cell Biol 183(3):377–384. doi:10.1083/jcb.200808137

Jonassen JA, SanAgustin J, Baker SP, Pazour GJ (2012) Disruption of IFT complex A causes cystic kidneys without mitotic spindle misorientation. J Am Soc Nephrol JASN 23(4):641–651. doi:10.1681/ASN.2011080829

Keady BT, Samtani R, Tobita K, Tsuchya M, San Agustin JT, Follit JA, Jonassen JA, Subramanian R, Lo CW, Pazour GJ (2012) IFT25 links the signal-dependent movement of Hedgehog components to intraflagellar transport. Dev Cell 22(5):940–951. doi:10.1016/j.devcel.2012.04.009

Keith DS, Torres VE, Johnson CM, Holley KE (1994) Effect of sodium chloride, enalapril, and losartan on the development of polycystic kidney disease in Han:SPRD rats. Am J Kidney Dis Off J Nat Kidney Found 24(3):491–498

Kim I, Ding T, Fu Y, Li C, Cui L, Li A, Lian P, Liang D, Wang DW, Guo C, Ma J, Zhao P, Coffey RJ, Zhan Q, Wu G (2009) Conditional mutation of Pkd2 causes cystogenesis and upregulates beta-catenin. J Am Soc Nephrol JASN 20(12):2556–2569. doi:10.1681/ASN.2009030271

Klawitter J, Zafar I, Klawitter J, Pennington AT, Klepacki J, Gitomer BY, Schrier RW, Christians U, Edelstein CL (2013) Effects of lovastatin treatment on the metabolic distributions in the Han:SPRD rat model of polycystic kidney disease. BMC Nephrol 14:165. doi:10.1186/1471-2369-14-165

Ko JY, Park JH (2013) Mouse models of polycystic kidney disease induced by defects of ciliary proteins. BMB Rep 46(2):73–79

Lehman JM, Michaud EJ, Schoeb TR, Aydin-Son Y, Miller M, Yoder BK (2008) The Oak Ridge Polycystic Kidney mouse: modeling ciliopathies of mice and men. Dev Dynam Off Publ Am Assoc Anat 237(8):1960–1971. doi:10.1002/dvdy.21515

Li X (2011) Epigenetics and autosomal dominant polycystic kidney disease. Biochim Biophys Acta 1812(10):1213–1218. doi:10.1016/j.bbadis.2010.10.008

Liu S, Lu W, Obara T, Kuida S, Lehoczky J, Dewar K, Drummond IA, Beier DR (2002) A defect in a novel Nek-family kinase causes cystic kidney disease in the mouse and in zebrafish. Development 129(24):5839–5846

Masyuk TV, Masyuk AI, Torres VE, Harris PC, Larusso NF (2007) Octreotide inhibits hepatic cystogenesis in a rodent model of polycystic liver disease by reducing cholangiocyte adenosine 3′,5′-cyclic monophosphate. Gastroenterology 132(3):1104–1116. doi:10.1053/j.gastro.2006.12.039

Meijer E, Drenth JP, D'Agnolo H, Casteleijn NF, de Fijter JW, Gevers TJ, Kappert P, Peters DJ, Salih M, Soonawala D, Spithoven EM, Torres VE, Visser FW, Wetzels JF, Zietse R, Gansevoort RT, Consortium D (2014) Rationale and design of the DIPAK 1 study: a randomized controlled clinical trial assessing the efficacy of lanreotide to halt disease progression in autosomal dominant polycystic kidney disease. Am J Kidney Dis Off J Nat Kidney Found 63(3):446–455. doi:10.1053/j.ajkd.2013.10.011

Miranda N, Miranda F, Rinaldi L, Stratigis S, Capasso G (2013) [Inhibitors of intra-cystic secretion: novel therapies in ADPKD (Autosomal Dominant Polycystic Kidney Disease)]. Giornale italiano di nefrologia : organo ufficiale della Societa italiana di nefrologia 30 (1)

Muto S, Kawano H, Higashihara E, Narita I, Ubara Y, Matsuzaki T, Ouyang J, Torres VE, Horie S (2015) The effect of tolvaptan on autosomal dominant polycystic kidney disease patients: a subgroup analysis of the Japanese patient subset from TEMPO 3:4 trial. Clin Exp Nephrol 19(5):867–877. doi:10.1007/s10157-015-1086-2

Nagao S, Yamaguchi T, Kusaka M, Maser RL, Takahashi H, Cowley BD, Grantham JJ (2003) Renal activation of extracellular signal-regulated kinase in rats with autosomal-dominant polycystic kidney disease. Kidney Int 63(2):427–437. doi:10.1046/j.1523-1755.2003.00755.x

Nagao S, Morita M, Kugita M, Yoshihara D, Yamaguchi T, Kurahashi H, Calvet JP, Wallace DP (2010) Polycystic kidney disease in Han:SPRD Cy rats is associated with elevated expression and mislocalization of SamCystin. Am J Physiol Renal Physiol 299(5):F1078–F1086. doi:10.1152/ajprenal.00504.2009

Nagao S, Kugita M, Yoshihara D, Yamaguchi T (2012) Animal models for human polycystic kidney disease. Exp Anim/Jpn Assoc Lab Anim Sci 61(5):477–488

Nakamura T, Ebihara I, Nagaoka I, Tomino Y, Nagao S, Takahashi H, Koide H (1993) Growth factor gene expression in kidney of murine polycystic kidney disease. J Am Soc Nephrol JASN 3(7):1378–1386

Natoli TA, Smith LA, Rogers KA, Wang B, Komarnitsky S, Budman Y, Belenky A, Bukanov NO, Dackowski WR, Husson H, Russo RJ, Shayman JA, Ledbetter SR, Leonard JP, Ibraghimov-Beskrovnaya O (2010) Inhibition of glucosylceramide accumulation results in effective blockade of polycystic kidney disease in mouse models. Nat Med 16(7):788–792. doi:10.1038/nm.2171

Novalic Z, van der Wal AM, Leonhard WN, Koehl G, Breuning MH, Geissler EK, de Heer E, Peters DJ (2012) Dose-dependent effects of sirolimus on mTOR signaling and polycystic kidney disease. J Am Soc Nephrol JASN 23(5):842–853. doi:10.1681/ASN.2011040340

Omori S, Hida M, Fujita H, Takahashi H, Tanimura S, Kohno M, Awazu M (2006) Extracellular signal-regulated kinase inhibition slows disease progression in mice with polycystic kidney disease. J Am Soc Nephrol JASN 17(6):1604–1614. doi:10.1681/ASN.2004090800

Park EY, Sung YH, Yang MH, Noh JY, Park SY, Lee TY, Yook YJ, Yoo KH, Roh KJ, Kim I, Hwang YH, Oh GT, Seong JK, Ahn C, Lee HW, Park JH (2009) Cyst formation in kidney via B-Raf signaling in the PKD2 transgenic mice. J Biol Chem 284(11):7214–7222. doi:10.1074/jbc.M805890200

Pazour GJ, Dickert BL, Vucica Y, Seeley ES, Rosenbaum JL, Witman GB, Cole DG (2000) Chlamydomonas IFT88 and its mouse homologue, polycystic kidney disease gene tg737, are required for assembly of cilia and flagella. J Cell Biol 151(3):709–718

Rocco MV, Neilson EG, Hoyer JR, Ziyadeh FN (1992) Attenuated expression of epithelial cell adhesion molecules in murine polycystic kidney disease. Am J Phys 262(4 Pt 2):F679–F686

Ruggenenti P, Remuzzi A, Ondei P, Fasolini G, Antiga L, Ene-Iordache B, Remuzzi G, Epstein FH (2005) Safety and efficacy of long-acting somatostatin treatment in autosomal-dominant polycystic kidney disease. Kidney Int 68(1):206–216. doi:10.1111/j.1523-1755.2005.00395.x

Sabers CJ, Martin MM, Brunn GJ, Williams JM, Dumont FJ, Wiederrecht G, Abraham RT (1995) Isolation of a protein target of the FKBP12-rapamycin complex in mammalian cells. J Biol Chem 270(2):815–822

Serra AL, Poster D, Kistler AD, Krauer F, Raina S, Young J, Rentsch KM, Spanaus KS, Senn O, Kristanto P, Scheffel H, Weishaupt D, Wuthrich RP (2010) Sirolimus and kidney growth in autosomal dominant polycystic kidney disease. N Engl J Med 363(9):820–829. doi:10.1056/NEJMoa0907419

Shibazaki S, Yu Z, Nishio S, Tian X, Thomson RB, Mitobe M, Louvi A, Velazquez H, Ishibe S, Cantley LG, Igarashi P, Somlo S (2008) Cyst formation and activation of the extracellular regulated kinase pathway after kidney specific inactivation of Pkd1. Hum Mol Genet 17(11):1505–1516. doi:10.1093/hmg/ddn039

Shillingford JM, Murcia NS, Larson CH, Low SH, Hedgepeth R, Brown N, Flask CA, Novick AC, Goldfarb DA, Kramer-Zucker A, Walz G, Piontek KB, Germino GG, Weimbs T (2006) The mTOR pathway is regulated by polycystin-1, and its inhibition reverses renal cystogenesis in polycystic kidney disease. Proc Natl Acad Sci U S A 103(14):5466–5471. doi:10.1073/pnas.0509694103

Shillingford JM, Piontek KB, Germino GG, Weimbs T (2010) Rapamycin ameliorates PKD resulting from conditional inactivation of Pkd1. J Am Soc Nephrol JASN 21(3):489–497. doi:10.1681/ASN.2009040421

Shillingford JM, Leamon CP, Vlahov IR, Weimbs T (2012) Folate-conjugated rapamycin slows progression of polycystic kidney disease. J Am Soc Nephrol JASN 23(10):1674–1681. doi:10.1681/ASN.2012040367

Smith LA, Bukanov NO, Husson H, Russo RJ, Barry TC, Taylor AL, Beier DR, Ibraghimov-Beskrovnaya O (2006) Development of polycystic kidney disease in juvenile cystic kidney mice: insights into pathogenesis, ciliary abnormalities, and common features with human disease. J Am Soc Nephrol JASN 17(10):2821–2831. doi:10.1681/ASN.2006020136

Sohara E, Luo Y, Zhang J, Manning DK, Beier DR, Zhou J (2008) Nek8 regulates the expression and localization of polycystin-1 and polycystin-2. J Am Soc Nephrol JASN 19(3):469–476. doi:10.1681/ASN.2006090985

Spirli C, Morell CM, Locatelli L, Okolicsanyi S, Ferrero C, Kim AK, Fabris L, Fiorotto R, Strazzabosco M (2012) Cyclic AMP/PKA-dependent paradoxical activation of Raf/MEK/ERK signaling in polycystin-2 defective mice treated with sorafenib. Hepatology 56(6):2363–2374. doi:10.1002/hep.25872

Sullivan LP, Wallace DP, Grantham JJ (1998) Chloride and fluid secretion in polycystic kidney disease. J Am Soc Nephrol JASN 9(5):903–916

Ta MH, Rao P, Korgaonkar M, Foster SF, Peduto A, Harris DC, Rangan GK (2014) Pyrrolidine dithiocarbamate reduces the progression of total kidney volume and cyst enlargement in experimental polycystic kidney disease. Physiol Rep 2(12):e12196. doi:10.14814/phy2.12196

Takiar V, Nishio S, Seo-Mayer P, King JD Jr, Li H, Zhang L, Karihaloo A, Hallows KR, Somlo S, Caplan MJ (2011) Activating AMP-activated protein kinase (AMPK) slows renal cystogenesis. Proc Natl Acad Sci U S A 108(6):2462–2467. doi:10.1073/pnas.1011498108

Tao Y, Kim J, Faubel S, Wu JC, Falk SA, Schrier RW, Edelstein CL (2005a) Caspase inhibition reduces tubular apoptosis and proliferation and slows disease progression in polycystic kidney disease. Proc Natl Acad Sci U S A 102(19):6954–6959. doi:10.1073/pnas.0408518102

Tao Y, Kim J, Stanley M, He Z, Faubel S, Schrier RW, Edelstein CL (2005b) Pathways of caspase-mediated apoptosis in autosomal-dominant polycystic kidney disease (ADPKD). Kidney Int 67(3):909–919. doi:10.1111/j.1523-1755.2005.00155.x

Torres VE, Wang X, Qian Q, Somlo S, Harris PC, Gattone VH 2nd (2004) Effective treatment of an orthologous model of autosomal dominant polycystic kidney disease. Nat Med 10(4):363–364. doi:10.1038/nm1004

Torres VE, Boletta A, Chapman A, Gattone V, Pei Y, Qian Q, Wallace DP, Weimbs T, Wuthrich RP (2010) Prospects for mTOR inhibitor use in patients with polycystic kidney disease and hamartomatous diseases. Clin J Am Soc Nephrol CJASN 5(7):1312–1329. doi:10.2215/CJN.01360210

Torres VE, Meijer E, Bae KT, Chapman AB, Devuyst O, Gansevoort RT, Grantham JJ, Higashihara E, Perrone RD, Krasa HB, Ouyang JJ, Czerwiec FS (2011) Rationale and design of the TEMPO (Tolvaptan Efficacy and Safety in Management of Autosomal Dominant Polycystic Kidney Disease and its Outcomes) 3–4 Study. Am J Kidney Dis Off J Nat Kidney Found 57(5):692–699. doi:10.1053/j.ajkd.2010.11.029

Torres VE, Chapman AB, Devuyst O, Gansevoort RT, Grantham JJ, Higashihara E, Perrone RD, Krasa HB, Ouyang J, Czerwiec FS, Investigators TT (2012) Tolvaptan in patients with autosomal dominant polycystic kidney disease. N Engl J Med 367(25):2407–2418. doi:10.1056/NEJMoa1205511

Wahl PR, Serra AL, Le Hir M, Molle KD, Hall MN, Wuthrich RP (2006) Inhibition of mTOR with sirolimus slows disease progression in Han:SPRD rats with autosomal dominant polycystic kidney disease (ADPKD). Nephrol Dial Transplant Off Publ Eur Dial Transplant Assoc Eur Renal Assoc 21(3):598–604. doi:10.1093/ndt/gfi181

Wallace DP, Rome LA, Sullivan LP, Grantham JJ (2001) cAMP-dependent fluid secretion in rat inner medullary collecting ducts. Am J Physiol Renal Physiol 280(6):F1019–F1029

Walz G, Budde K, Mannaa M, Nurnberger J, Wanner C, Sommerer C, Kunzendorf U, Banas B, Horl WH, Obermuller N, Arns W, Pavenstadt H, Gaedeke J, Buchert M, May C, Gschaidmeier H, Kramer S, Eckardt KU (2010) Everolimus in patients with autosomal dominant polycystic kidney disease. N Engl J Med 363(9):830–840. doi:10.1056/NEJMoa1003491

Wander SA, Hennessy BT, Slingerland JM (2011) Next-generation mTOR inhibitors in clinical oncology: how pathway complexity informs therapeutic strategy. J Clin Invest 121(4):1231–1241. doi:10.1172/JCI44145

Wang X, Wu Y, Ward CJ, Harris PC, Torres VE (2008) Vasopressin directly regulates cyst growth in polycystic kidney disease. J Am Soc Nephrol JASN 19(1):102–108. doi:10.1681/ASN.2007060688

Warner G, Hein KZ, Nin V, Edwards M, Chini CC, Hopp K, Harris PC, Torres VE, Chini EN (2015) Food restriction ameliorates the development of polycystic kidney disease. J Am Soc Nephrol JASN. doi:10.1681/ASN.2015020132

Watnick T, Germino GG (2010) mTOR inhibitors in polycystic kidney disease. N Engl J Med 363(9):879–881. doi:10.1056/NEJMe1006925

Wu G, Markowitz GS, Li L, D'Agati VD, Factor SM, Geng L, Tibara S, Tuchman J, Cai Y, Park JH, van Adelsberg J, Hou H Jr, Kucherlapati R, Edelmann W, Somlo S (2000) Cardiac defects and renal failure in mice with targeted mutations in Pkd2. Nat Genet 24(1):75–78. doi:10.1038/71724

Yamaguchi T, Nagao S, Wallace DP, Belibi FA, Cowley BD, Pelling JC, Grantham JJ (2003) Cyclic AMP activates B-Raf and ERK in cyst epithelial cells from autosomal-dominant polycystic kidneys. Kidney Int 63(6):1983–1994. doi:10.1046/j.1523-1755.2003.00023.x

Yamaguchi T, Reif GA, Calvet JP, Wallace DP (2010) Sorafenib inhibits cAMP-dependent ERK activation, cell proliferation, and in vitro cyst growth of human ADPKD cyst epithelial cells. Am J Physiol Renal Physiol 299(5):F944–F951. doi:10.1152/ajprenal.00387.2010

Yang B, Sonawane ND, Zhao D, Somlo S, Verkman AS (2008) Small-molecule CFTR inhibitors slow cyst growth in polycystic kidney disease. J Am Soc Nephrol JASN 19(7):1300–1310. doi:10.1681/ASN.2007070828

Ye M, Grant M, Sharma M, Elzinga L, Swan S, Torres VE, Grantham JJ (1992) Cyst fluid from human autosomal dominant polycystic kidneys promotes cyst formation and expansion by renal epithelial cells in vitro. J Am Soc Nephrol JASN 3(4):984–994

Yuajit C, Chatsudthipong V (2016) Nutraceutical for autosomal dominant polycystic kidney disease therapy. J Med Assoc Thailand Chotmaihet thangphaet 99(Suppl 1):S97–S103

Yuajit C, Muanprasat C, Gallagher AR, Fedeles SV, Kittayaruksakul S, Homvisasevongsa S, Somlo S, Chatsudthipong V (2014) Steviol retards renal cyst growth through reduction of CFTR expression and inhibition of epithelial cell proliferation in a mouse model of polycystic kidney disease. Biochem Pharmacol 88(3):412–421. doi:10.1016/j.bcp.2014.01.038

Zafar I, Tao Y, Falk S, McFann K, Schrier RW, Edelstein CL (2007) Effect of statin and angiotensin-converting enzyme inhibition on structural and hemodynamic alterations in autosomal dominant polycystic kidney disease model. Am J Physiol Renal Physiol 293(3):F854–F859. doi:10.1152/ajprenal.00059.2007

Chapter 8
Diagnostic Evaluation as a Biomarker in Patients with ADPKD

Hayne Cho Park and Curie Ahn

Abstract Recently, newer treatments have been introduced for autosomal dominant polycystic kidney disease (ADPKD) patients. Since cysts grow and renal function declines over a long period of time, the evaluation of treatment effects in ADPKD has been very difficult. Therefore, there has been a great interest to find out the "better" surrogate marker or biomarker which reflects disease progression. Biomarkers in ADPKD should have three clinical implications: (1) They should reflect disease severity, (2) they should distinguish patients with poor versus good prognosis to select those who will benefit better from the treatment, and (3) they should be easy to evaluate short-term outcome after treatment, which will demonstrate hard outcome. Herein, we will discuss currently available surrogate biomarkers including the volume of total kidney and urinary molecular markers.

Keywords Polycystic kidney, autosomal dominant • Biological markers • Patient selection • Prognosis • Treatment outcome

8.1 Introduction

Autosomal dominant polycystic kidney disease (ADPKD) is the most common hereditary kidney disease, which is characterized by progressive multiple cyst formation, proliferation, and apoptosis, finally causing interstitial fibrosis and end-stage renal disease (ESRD) (Torres et al. 2007). It is caused by genetic mutation on either *PKD1* or *PKD2* gene; however, it is not a congenital disease since phenotypic

H.C. Park (✉)
Division of Nephrology, Department of Internal Medicine, The Armed Forces Capital Hospital, Seongnam-si, Gyeonggi-do, South Korea
e-mail: haynepark798@gmail.com

C. Ahn
Department of Internal Medicine, Seoul National University College of Medicine, Seoul, South Korea
e-mail: curie@snu.ac.kr

J.H. Park, C. Ahn (eds.), *Cystogenesis*, Advances in Experimental Medicine and Biology 933, DOI 10.1007/978-981-10-2041-4_8

presentation only begins in young adulthood. Furthermore, renal function remains stable at the early stage because glomerular hyperfiltration compensates for the progressive loss of healthy glomeruli, which result in renal failure only after several decades from diagnosis. Interestingly, even within a family, individual members show different prognosis. Therefore, it is not a pure 'genetic' disease but a 'genetic-environmental' disease.

Recently, several novel therapies have been introduced to slow the rate of disease progression in ADPKD. Since ADPKD progresses slowly after cyst formation, typical hard outcomes such as time to ESRD or patient survival are often not useful in studying ADPKD treatment effect. Therefore, there has been a great interest to develop alternative endpoints or surrogate biomarkers for patient and renal outcomes. Surrogate biomarkers should fulfill following conditions. Firstly, they should reflect disease severity. Second, they should predict rapid disease progression, which will distinguish high-risk patients who will benefit from the treatment. Lastly, they should be easy and convenient to evaluate disease progression in a short-term interval. Herein, we will discuss diagnostic evaluation of currently available surrogate biomarkers in the patients with ADPKD.

8.2 Biomarker as a Predictor of Rapid Disease Progression

Since renal function declines only at the late stage of the disease, early biomarkers to predict renal disease progression need to be clarified. This part will seek for markers used to assess prognosis in ADPKD (Table 8.1).

Table 8.1 Biomarkers as a predictor of rapid disease progression

Genetic factor	*PKD1* >> *PKD2*
	Truncating mutation in *PKD1* >> In-frame mutation >> Missense mutation
	Pathologic mutation on second allele (Hypomorphic allele)
	Modifier gene
	Hypermethylation on *PKD1*
Demographic factors	Young age at diagnosis
	Male >> Female
	Family history of early ESRD < 55years
Clinical factors	Height-adjusted TKV ≥ 600 mL/m
	Early onset of hypertension < 30years
	Severe hypertension
	Previous history of gross hematuria
	Proteinuria
	Recurrent urinary tract infection or pyuria

8.2.1 *Genetic Factors*

ADPKD is hereditary disease which involves mutations in *PKD1* and/or *PKD2*. Recent advances in molecular diagnosis have led us to more precisely predict the renal disease progression in ADPKD patients. Not only type of genes but also type or location of mutations would make a difference in prognosis. We will discuss current understanding of genetic prognostic factors.

8.2.1.1 Locus Heterogeneity

ADPKD is genetically heterogeneous, in which cystic phenotype is caused by mutations in genes at different chromosomal loci: *PKD1* (chromosome 16p 13.3) and *PKD2* (chromosome 4q21). This is called locus heterogeneity. Previously, a possibility of third gene locus has been proposed and searched since five pedigrees could not find pathologic mutations from either *PKD1* or *PKD2* by linkage analysis (Paterson and Pei 1998). However, reanalysis of clinical data and genetic samples from '*PKD3* pedigree' by mutation screening revealed that '*PKD3*' is unlikely to exist (Paul et al. 2014). Therefore, currently it is widely accepted that only *PKD1* and *PKD2* genes are responsible for ADPKD development.

Patients with *PKD1* mutations generally show a more severe form of ADPKD than those with *PKD2* mutation, showing earlier onset of hypertension, earlier onset of diagnosis, larger kidneys, faster renal function decline, and earlier onset of ESRD. In the European *PKD1-PKD2* cohort study, *PKD1* pedigree showed earlier onset of ESRD or death compared to *PKD2* pedigree (53.0 years vs. 69.1 years) (Hateboer et al. 1999). The incidence of other clinical manifestations such as hypertension, hematuria or urinary tract infection was also less frequent in *PKD2* pedigree. Similarly, patients with *PKD1* mutation have more cysts in kidneys resulting in larger kidney size compared to patients with *PKD2* mutation (Harris et al. 2006).

8.2.1.2 Allelic Heterogeneity

The *PKD1* and *PKD2* genes present a high level of allelic heterogeneity. Within each gene, different type and location of mutations can cause cyst formation. These variable mutations are collected and reported in the ADPKD Mutation Database (PKDB, http://pkdb.mayo.edu/). To date, 1273 pathogenic mutations in *PKD1* and 202 pathogenic mutations in *PKD2* have been described in the database. Recently, the role of allelic heterogeneity in modulating disease severity has been studied. The Consortium for Radiologic Imaging Study of PKD (CRISP) researchers found that missense mutation in *PKD1* presents with mild phenotype similar to *PKD2* phenotype (Pei et al. 2012). Another study from Genkyst cohort compared renal survival among 387 patients with *PKD1* truncating mutation, 184 patients with *PKD1* non-truncating mutation, and 133 patients with *PKD2* mutation (Cornec-Le Gall et al.

2013). As the result, the patients with *PKD1* non-truncating mutation (in-frame mutation or missense mutation) demonstrated 12-year delayed onset of ESRD compared to *PKD1* truncating mutation (68 years vs. 56 years, $p < 0.0001$). Interestingly, recent study suggested that *PKD1* non-truncating mutations can show heterogeneous renal prognosis according to gene dosage (Hwang et al. 2015). For example, some *PKD1* missense mutations result in milder form of disease similar to *PKD2* mutation, while other *PKD1* missense mutations can show deleterious outcome such as *PKD1* truncating mutation. The allelic effect according to mutation location within *PKD1* non-truncating mutations should be elucidated further.

8.2.1.3 Hypomorphic Allele

A mutation that reduces but doesn't eliminate a gene's functionality is hypomorphic. Hypomorphic alleles in ADPKD can be demonstrated as a mild phenotype, but they can result in a severe disease when they occur together with a pathogenic mutation on the second allele (Rossetti et al. 2009; Bergmann et al. 2011). This phenomenon can explain why some members from the family of mild phenotype show extremely severe disease form. Segregation study may be useful to find out hypomorphic allele effect within the family (Hwang et al. 2015).

8.2.1.4 Modifier Genes

Intra-familial variability is another feature of ADPKD. Locus heterogeneity and allelic heterogeneity can explain variable disease severity among different pedigrees. However, within the same pedigree, family members may show variable renal disease progression (Barua et al. 2009). This can be partially explained with modifier gene effect. Several gene polymorphisms such as angiotensin converting enzyme (ACE) (Perez-Oller et al. 1999), Dickkopf 3 (DKK3) (Liu et al. 2010), endothelial nitric oxide synthase (Persu et al. 2002) may be associated with poorer prognosis.

8.2.1.5 Epigenetic Modification

Recently, epigenetic modification effect on ADPKD phenotype has become focus of scientific research. Epigenetic regulation is defined as the modification of external or environmental factors that switch genes on and off and affect how cells read genes instead of changing DNA sequence (Park et al. 2011). Epigenetic modulation includes DNA methylation, histone modification, and gene modification by microRNAs. ADPKD patients showed hyper-methylation on *PKD1* gene and other cystogenesis-related genes and subsequent downregulation of *PKD1* expression (Woo et al. 2014). Moreover, histone deacetylases (HDACs) are known to be activated in ADPKD kidneys, which result in dysregulated *PKD1* gene expression, disrupted fluid-flow regulated calcium signal regulation in renal epithelia, and subsequent cyst formation and progression (Li 2011). Recently, there is increasing evidence that microRNA dysregulation is associated with the pathogenesis of cystic kidney disease (Woo and Park 2013; Lakhia et al. 2015; Sun et al. 2015) and urinary microRNAs can be a potential biomarker for ADPKD progression (Ben-Dov et al. 2014). Therefore, epigenetic modification is discussed as a possible therapeutic target of ADPKD.

8.2.2 *Demographic and Clinical Factors*

Apart from the culprit gene affecting ADPKD, non-modifiable and modifiable factors affect ADPKD renal disease progression. Since genetic analysis is not the routine diagnostic work up at the clinic, collecting demographic and clinical factors and family history would be most helpful to determine prognosis and to counsel each patient.

8.2.2.1 Age at Diagnosis

Children with early severe disease showed faster renal volume growth and faster decline of renal function on follow up (Fick-Brosnahan et al. 2001). Another study showed that children who were diagnosed with PKD<18 months demonstrated larger kidneys and severe hypertension compared to those who were diagnosed after 18 months of age (Shamshirsaz et al. 2005). In a similar vein, patients with *PKD1* genotype showed more cysts than *PKD2* genotype, but growth rates were not different between two genotypes (Harris et al. 2006). Altogether, these results suggest that diagnosis of ADPKD at the early age is likely to mean severe genotype.

8.2.2.2 Male Gender

Men is known to have faster renal function decline, larger kidneys, and hypertension in ADPKD resulting in earlier onset of ESRD. Gabow et al. described that men had a faster decline of renal function compared to female counterpart (Gabow et al. 1992b). Another retrospective study of 1215 ADPKD patients, male gender showed earlier onset of ESRD compared to female gender (52 vs. 56 years) (Johnson and Gabow 1997).

8.2.2.3 Family History of ESRD

As we described above, genetic factor is the strongest determinant of renal prognosis. The problem is that we cannot perform genetic analysis in every case due to practical and economic issue. One study group performed the interesting study to show that family history of ESRD onset can predict mutated gene (Barua et al. 2009). In this study, they examined 484 affected members from 90 pedigrees whose pathogenic mutations are already known. The researcher found that we can predict gene locus by collecting the information of ESRD onset from family members. If any affected member shows ESRD onset<55 years, this family is highly likely to have *PKD1* gene mutation (positive predictive value 100 %, sensitivity 72 %). On the other hand, if none of family members experience ESRD until 70 years old, this family is highly likely to have PKD2 gene mutation (positive predictive value 100 %, sensitivity 74 %). This study emphasizes the importance of history taking and collecting family history to predict prognosis.

8.2.2.4 Early Onset and Severity of Hypertension

Elevated blood pressure is the most common comorbidity in ADPKD patients. It is the most common early manifestation and also the major contributor to renal disease progression. Hypertension is also associated with other risk factors such as hematuria, proteinuria, and cardiovascular outcomes. Intrarenal renin-angiotensin system (RAS) has been suggested as the major contributor to blood pressure elevation in ADPKD (Schrier 2011). Of interest, primary hypertension or essential hypertension in unaffected parent is also related to renal progression and early onset of hypertension in ADPKD offsprings (Geberth et al. 1995; Schrier et al. 2003). Therefore, there is a debate whether hypertension and related biomarkers are specific to ADPKD or they are general risk factors for chronic kidney disease progression.

Nonetheless, ADPKD patients with early onset hypertension demonstrate poorer renal outcome. One survival analysis demonstrated that patients with early hypertension<35 years showed faster ESRD progression compared to those with late onset hypertension (ESRD onset 51 vs. 65 years) (Johnson and Gabow 1997). In addition, severity of hypertension is also related to renal progression rate. Recent large scale, randomized prospective study showed that low blood pressure target (95/60 to 110/75 mmHg) group demonstrated attenuated renal cyst growth, reduced albuminuria, reduced left ventricular mass index compared to standard blood pressure target (120/70 to 130/80 mmHg) group (Schrier et al. 2014). Therefore, early onset hypertension and severity of hypertension both contribute as predictors of rapid disease progression.

8.2.2.5 Gross Hematuria

Gross hematuria may occur in as much as 40 % of ADPKD patients at some time during the course of the disease. Many of them can experience recurrent episode of gross hematuria or cyst rupture. A retrospective study demonstrated that patients with gross hematuria had larger kidneys and those with recurrent gross hematuria tended to have faster renal function decline (Gabow et al. 1992a). Another survival analysis showed that patients with gross hematuria before age 30 had a worse renal outcome compared to those with later onset (Idrizi et al. 2009). It was postulated that cyst rupture and subsequent gross hematuria may be the sign of faster growth of cyst and related to poor outcome (Kistler et al. 2009b). Others also claim that gross hematuria may release and deposit iron and heme promoting inflammation around kidney tissue (Tracz et al. 2007).

8.2.2.6 Proteinuria

Overt proteinuria (>300 mg/day) was associated with faster decline of renal function and early onset of ESRD in several studies. In prospective cohort study of 323 ADPKD patients with mean follow up duration of 8 ~ 9years, baseline proteinuria was significantly correlated with the declining rate of renal function (Ozkok et al. 2013). Secondary subgroup analysis from Modification of Diet in Renal Disease (MDRD) study revealed that higher degree of proteinuria was associated with faster decline of renal function (Klahr et al. 1995). Microalbuminuria (30 ~ 300 mg/day) also was strongly associated with larger kidneys faster growth of renal volume, and higher blood pressure (Kistler et al. 2009b). Whether proteinuria is the result of cyst growth and subsequent renal tissue injury of the cause of renal progression is not clearly defined at this moment.

8.2.2.7 Recurrent Urinary Tract Infection or Pyuria

Retrospective studies showed that urinary tract infection (UTI) is very common among ADPKD patients and female was more prone to overt UTI (Idrizi et al. 2011). Recent study also showed that asymptomatic pyuria is prevalent in ADPKD patients and chronic recurrent pyuria itself is associated with faster decline of renal function (Hwang et al. 2013). However, this single-center retrospective study has a major limitation in which culture study was not performed in most cases of pyuria. Pyuria as a potential biomarker should be elucidated further in prospective studies.

8.3 Biomarker as a Tracer of Disease Progression

ADPKD has a typically long period of preserved renal function at the early stage and renal function only begins to decline after kidneys are filled with enormous cysts. Moreover, at the early stage of disease, glomerular hyperfiltration compensates for the progressive loss of healthy glomeruli. Therefore, traditional markers such as serum creatinine or glomerular filtration rate (GFR) do not always reflect the disease progression. Hence, the following surrogate biomarkers to identify and trace the disease progression have been introduced (Table 8.2).

Table 8.2 Biomarkers as a tracer of disease progression

Annual decline rate of mGFR or eGFR		
Annual increase in TKV		
Novel biomarkers	Biomarkers of acute tubular injury	NGAL, NAG, KIM-1, β2-microglobulin, H-FABP
	Biomarkers of inflammation	Urinary MCP-1, plasma ADMA, Urinary complement 3 and 9, Urinary plakin
	Biomarkers of fibrosis	Urinary apelin, Urinary TGF-β1
	Biomarkers of hypertension	Urinary angiotensinogen, Serum uric acid
	Biomarkers of cAMP pathway	Plasma and urinary copeptin
	Biomarkers of Wnt pathway	Serum sFRP4

8.3.1 Decline Rate of GFR

GFR describes the flow rate of filtered fluid through the kidney and presents renal function. It is used instead of blood urea nitrogen or serum creatinine because plasma concentration of these waste substances will not be raised above the normal range until kidney function falls below 60 % of total kidney function. GFR can be measured by calculating the clearance of exogenous materials such as inulin or iothalamate which are neither reabsorbed nor secreted by the kidney after glomerular filtration. Since it is not practical to measure GFR at each clinic visit, estimated GFR (eGFR) is calculated by equation. There is still debates whether eGFR can be used instead of measured GFR (mGFR) as a tracer of renal progression in clinical trials. Early studies showed that mGFR may be superior to eGFR early in chronic kidney disease (CKD) (Rule et al. 2006; Ruggenenti et al. 2012). The eGFR measured by MDRD or Cockcroft-Gault equation may underestimate GFR in cohort with normal renal function > 60 mL/min/1.73 m^2. However, recent studies suggested that eGFR may perform relatively well compared to mGFR and reliably reflect GFR change in the patients with ADPKD (Spithoven et al. 2013; Chapman et al. 2015). Among eGFR equation, CKD-EPI equation should be considered then the trial is designed to include patients with normal renal function (Orskov et al. 2010).

Since eGFR measurement is not as accurate as mGFR, small changes in eGFR may not reflect a true decline in renal function. Therefore, data of repeated measurement of eGFR over a long period should be available to calculate the decline rate of eGFR. Recent review paper suggested to define rapid progressor as a confirmed annual eGFR decline ≥ 5 mL/min/1.73 m^2/year or by an average annual eGFR decline ≥ 2.5 mL/min/1.73 m^2/year over 5 years (Gansevoort et al. 2016). However, absolute slopes (mL/min/year or mL/min/1.73 m^2/year) may not accurately reflect the true decline rate of eGFR. For example, a rise in serum creatinine from 0.9 to 1.0 mg/dL or from 3.3 to 6.0 mg/dL represents a decline in eGFR of 10 mL/min/1.73 m^2

for a 60-year old man by MDRD equation. In another words, small increase in serum creatinine may be interpreted as a large decrement of eGFR in early CKD. Therefore, logarithmic slopes (% per year) rather than absolute slopes (mL/min/1/73 m^2/year) should be considered in the early CKD stages.

8.3.2 *Total Kidney Volume*

Cysts develop as early as in utero but renal function declines very slowly thereafter until kidney fails in fifth or sixth decades. Therefore, GFR measurement often do not reflect disease progression in the early stage of the disease. Measurement of total kidney volume or cyst volume is considered excellent surrogate marker to reflect cyst growth and to predict further disease progression.

Renal volume can be measured either by ultrasonography, computed tomography (CT), or magnetic resonance imaging (MRI). Ultrasonography is an easy and safe screening method to diagnose and evaluate kidney volume in ADPKD. One can calculate renal volume by ellipsoid formula: height × transverse width × thickness × π/6. However, it is an operator-dependent, less reproducible, and underestimates the true value. Therefore, it is not suitable for short-term follow up of renal progression. On the other hands, CT and MRI can detect as small cyst as 2 mm, reliably measure volume progression in a short interval, and are highly reproducible (Bae and Grantham 2010). Therefore, they are more suitable as surrogate markers for clinical trials. Similar to eGFR, kidney volume can be estimated by formula instead of measured TKV (mTKV). Recently, some researchers suggested that estimated TKV (eTKV) by ellipsoid formula or midslice method can be conveniently and reliably used instead of mTKV (Spithoven et al. 2015). Therefore, both mTKV and eTKV can be used as a tracer for disease progression in ADPKD.

However, there can still be intra-observer and intra-individual variability in renal volume measurement. Therefore, it is recommended to at least measure renal volume more than three times with 6–12 month apart to assess volume change (Gansevoort et al. 2016). It is well documented in the previous study that mTKV by MRI imaging can detect cyst growth within 6 months (Kistler et al. 2009b). When previous data or repeated measurements of renal volume are not available, however, we can still make prediction by single TKV value. Chapman et al. showed that height-adjusted TKV (HtTKV) ≥ 600 mL/m predict the risk of developing CKD stage 3 within 8 years (Chapman et al. 2012). Irazabal et al. developed risk prediction tool from a cohort of 590 ADPKD patients and classified them into five groups according to HtTKV ranges for age (1A-1E, in the order of increasing risk of ESRD development) (Irazabal et al. 2015). The risk of ESRD development within 10 years was increased from 2.4 to 66.9 % according to subclasses (1A vs. 1E). In summary, single and repeated mTKV or eTKV can be used as a useful surrogate biomarker for renal disease progression in ADPKD.

8.3.3 Serum and Urinary Biomarkers

8.3.3.1 General Biomarkers of Kidney Disease

ADPKD is initiated from cystic transformation of a few percent of nephrons leading to increased cell proliferation and apoptosis, interstitial inflammation, fibrosis, and finally ESRD. Therefore, ADPKD kidney may show the whole spectrum of acute kidney injury and chronic kidney disease (CKD) along the disease course. Hence, various serum and urinary biomarkers of kidney disease have been studies as a surrogate marker of disease progression. Since serum creatinine based GFR is considered of limited use and measurement of TKV may be time-consuming and expensive, serum and urinary biomarkers can be easy and inexpensive method to trace disease progression in the future.

Biomarkers of Acute Tubular Damage

Since cysts initiate from tubules, the potential markers of tubular damage were studied widely to seek the relationship with renal function change. In one cross-sectional study with 107 ADPKD patients from CRISP cohort, urinary neutrophil gelatinase-associated lipocalin (NGAL) and Interleukin-18 (Il-18) were stably increased but was not associated with either TKV or eGFR (Parikh et al. 2012). Another cross-sectional study revealed that markers of proximal tubular damage (NGAL, β2-microglobulin) and the marker of distal tubular damage (heart-type fatty acid binding protein, H-FABP) were negatively associated with mGFR. NGAL was also positively associated with TKV together with kidney injury molecule 1 (KIM-1) (Meijer et al. 2010). However, whether these potential tubular damage markers can predict disease progression needs to be further elucidated. One prospective study showed that N-acetyl-β-D glucosaminidase (NAG) is better associated with eGFR compared to KIM-1, β2-microglobulin, and NGAL. However, urinary NAG was failed to predict renal function decline in 1 year (Park et al. 2012).

Biomarkers of Inflammation

Inflammation and oxidative stress is evident from early stage of disease. Urinary biomarkers to measure intrarenal inflammation and oxidative stress have been studied. Previous cross-sectional studies showed that monocyte chemoattractant protein-1 (MCP-1) was elevated in the urine of ADPKD patients and it was associated with decreased renal function (Zheng et al. 2003; Parikh et al. 2012). Another study demonstrated that the level of asymmetric dimethylarginine (ADMA), an inhibitor of nitric oxide synthase, is increased in patients with early ADPKD (Wang et al. 2008). Most recently, proteomic analysis of urinary vesicle revealed that the level of complements 3 and 9 (C3 and C9) were elevated in the early stage of ADPKD and urinary plakins were elevated in the later stage of the disease (Salih et al. 2016). Both complements and plakin levels were associated with TKV. This study suggests that inflammation process takes an important role in disease progression in the early stage of ADPKD.

Biomarkers of Fibrosis

ADPKD kidneys eventually change to fibrotic kidneys similar to other types of CKD. However, unlike the CKD of other causes, fibrosis starts from the epithelial change from cyst formation, which drives changes in the peri-cystic interstitium and fibroblasts (Norman 2011). Previous studies tried to discover urinary biomarker reflecting the early fibrotic process in the polycystic kidneys. One researcher group performed proteomic analyses showing that collagen fragments take most part of excreted peptide in the urine (Kistler et al. 2009a, 2013). The results suggested that fibrotic process takes the important role in ADPKD progression.

Apelin, an endogenous ligand of the G-protein-coupled receptor APJ, has been recently identified as a main regulator of organ fibrosis. In kidneys, Apelin-APJ axis has been known to protect kidneys from fibrosis (Huang et al. 2016). One recent study compared the excretion levels of fibrotic markers, apelin and transforming growth factor-β1 (TGF-β1), in the urine samples from 45 ADPKD patients and 28 healthy controls (Kocer et al. 2016). The results showed that urinary apelin level was lower and TGF-β1 level was higher in ADPKD cohort compared to the healthy control. In one prospective cohort study of 52 ADPKD patients, apelin has been investigated for its predictive value (Lacquaniti et al. 2013). Of 52 patients, 33 patients reached primary outcome (combined measurement of decreased GFR and increased TKV by 5 % per year). Apelin independently predicted the renal progression in ADPKD patients. Apelin and its role as a biomarker should be evaluated in a larger cohort with longer period.

Biomarkers of Hypertension

High blood pressure is one of the early complications of ADPKD. It is well known that intrarenal RAS contributes to the rise of blood pressure in ADPKD (Schrier 2011). Because blood pressure begins to rise as early as the stage with a few cyst, biomarkers that reflect intrarenal RAS activity and hypertension have been studied.

Urinary angiotensinogen (AGT) has been suggested as an effective biomarker to reflect intrarenal RAS activity and hypertension (Kobori et al. 2009). Since AGT cannot be filtered through glomeruli, urinary AGT solely reflect intrarenal RAS activity. In addition, urinary AGT is well known to be strongly associated with the activity of angiotensin-II, a main player in intrarenal RAS. In our previous study, urinary AGT was associated with eGFR and TKV. Hypertensive patients showed higher level of urinary AGT compared to normotensive patients (Park et al. 2015).

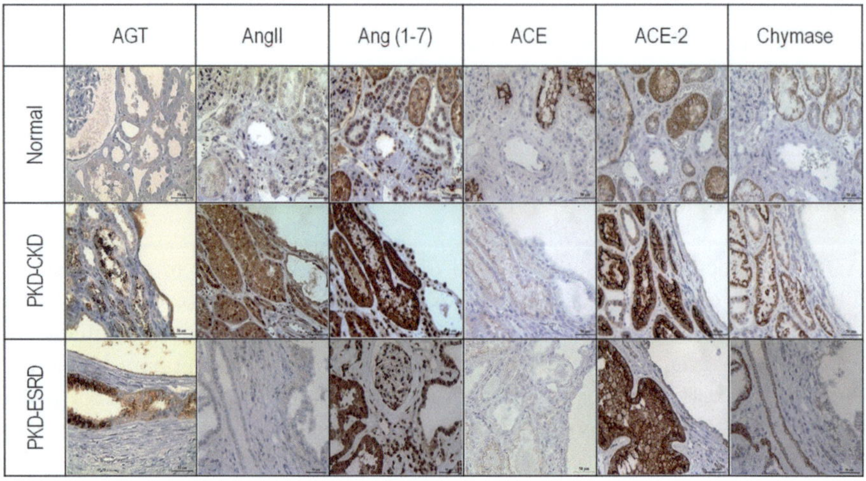

Fig. 8.1 Components of intrarenal RAS in polycystic kidneys (Reproduced from Park et al. 2015)

In addition, polycystic kidneys showed strong expression of AGT along the cyst-lining epithelium as early as CKD stage II (Fig. 8.1). However, the study failed to show the association of urinary AGT with decline rate of GFR or ESRD progression.

On the other hands, serum uric acid has recently been recognized as a marker of endothelial dysfunction, hypertension, and predictor of renal disease progression in ADPKD. One cross-sectional study evaluated the association of serum uric acid and endothelial dysfunction in early normotensive ADPKD patients (Kocyigit et al. 2013). Endothelial dysfunction was measured by flow-mediated vasodilation (FMD) in the forearm. The group of higher serum uric acid (>7.0 for males and >6.0 for females) showed higher ADMA levels (1.19 ± 0.2 vs. 1.47 ± 0.3, $p < 0.001$) and lower FMD rates (8.1 ± 1.3 vs. 6.8 ± 0.7, $p < 0.001$). Another retrospective analysis of a prospective cohort showed that higher serum uric acid level was associated with hypertension development, larger TKV, and risk of developing ESRD (Helal et al. 2013). This study demonstrated that the risk of ESRD increases in the 4th and 3rd quartiles of uric acid compared with the 1st [4.8 (2.6–8.9; $p < 0.001$)] and 2.9 [(1.6–5.3; $p < 0.001$)].

8.3.3.2 Biomarkers Reflecting Specific Pathway Activity

Biomarker of cAMP Pathway: Copeptin

Copeptin is the carboxy-terminal portion of AVP and serves as a marker for endogenous AVP levels. Because AVP activates cAMP pathway to stimulate cyst growth, plasma and urinary copeptin may be a valuable marker for renal disease progression. In a cross-sectional study, plasma copeptin level was associated with mGFR, albuminuria, and TKV (Meijer et al. 2011). In the following prospective cohort study, plasma level of copeptin was not only associated with baseline eGFR but also with decline rate of eGFR during follow up duration of mean 11.2 years (standard B −0.345, $p<0.01$) (Boertien et al. 2012). In addition, 8 out of 9 patients who developed ESRD during follow up showed higher than median value of plasma copeptin at the baseline. In a longitudinal observational study of CRISP cohort, plasma copeptin concentration was independently associated with change in TKV ($p>0.001$) and mGFR ($p=0.09$) (Boertien et al. 2013). Urinary copeptin was also investigated recently and researchers found that urinary copeptin was positively associated with TKV and negatively associated with eGFR (Nakajima et al. 2015). The predictive role of urinary copeptin should be further elucidated in the future study.

Biomarker of Wnt Pathway: sFRP4

The abnormal Wnt signaling has been suggested to have a role in the pathogenesis of ADPKD. Secreted Frizzled-related protein 4 (sFRP4) is a secreted molecule that antagonizes Wnt signaling pathway. The sFRP4 was well documented to be upregulated in human ADPKD and in 4 different PKD animal models (Romaker et al. 2009). Cyst fluid from ADPKD kidneys activated the sFRP4 production in renal tubular epithelial cell lines. Recent study showed that higher serum level of sFRP4 at the baseline may predict renal function decline in 24 months (Zschiedrich et al. 2016). The patients with sFRP4 level below 5 ng/mL at baseline experienced an average eGFR decline of 3.2/mL/min/1.73 m^2/year while those with sFRP4 level above 30 ng/mL at baseline experienced an average eGFR decline of 4.5 mL/min/1.73 m^2/year ($p=0.0063$). Whether urinary sFRP4 can predict disease progression in ADPKD needs to be further elucidated.

8.4 Identifying High-Risk Patients for Treatment

Not all the patients diagnosed with ADPKD reach end-stage. The variable disease courses make clinicians difficult to select the treatment boundary or guideline. Before several therapeutic agents became available, the goal of ADPKD management was similar to that of CKD – lowering blood pressure and treating complications. However, recent advances in molecular diagnosis and treatment strategy made us available to slow down the rate of disease progression. Therefore, selecting patients who are most likely to benefit from novel treatment should be considered before launching clinical trials.

Recent review paper well summarized the algorithm to select high-risk patients or rapid progressors for clinical trials (Gansevoort et al. 2016) (Fig. 8.2). First of all,

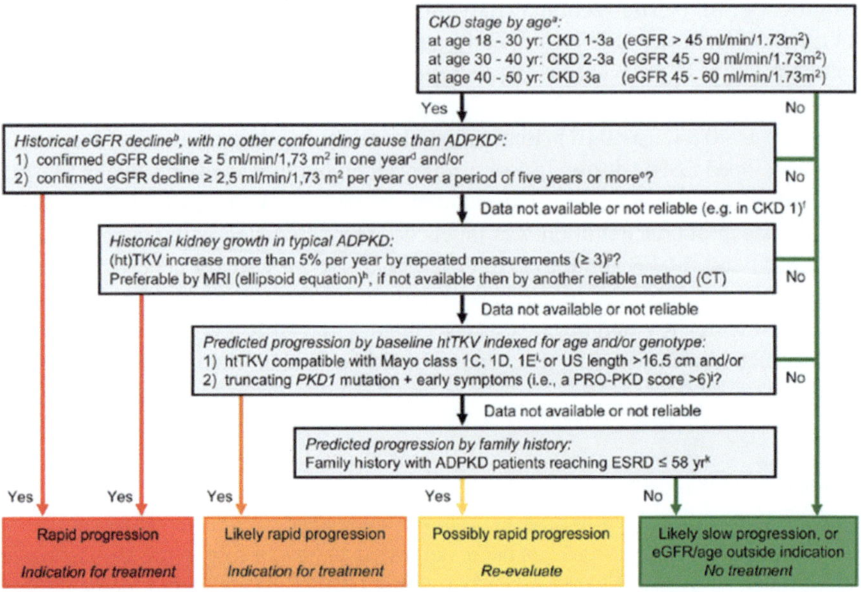

Fig. 8.2 Algorithm to assess indications for initiation of treatment in ADPKD (Reproduced from Gansevoort et al. 2016)

clinicians should consider baseline age and renal function before prescribing novel treatment. For example, a middle aged ADPKD patient in CKD stage 1 is less likely to progress rapidly in his life time. In the same manner, *PKD2* family member with preserved renal function is less likely to progress in a few years. Therefore, we should evaluate baseline age and renal function before patient selection. Next, we should define rapid progressors according to previous mentioned risk factors – decline rate of eGFR, TKV change in recent years, baseline TKV, PKD genotype, and family history of ESRD. By following this algorithm, we can expect the optimal outcome with minimal harm to the ADPKD subjects.

8.5 Conclusions

On the basis of improved knowledge in molecular pathophysiology, a large number of novel therapies have been proposed for ADPKD patients. As a clinician, selecting the patients who will benefit the most and monitoring disease progression are as important as initiating treatment. Various biomarkers described in this chapter will help us to define rapid progressor and to trace disease progression to get the best result from novel treatment.

References

Bae KT, Grantham JJ (2010) Imaging for the prognosis of autosomal dominant polycystic kidney disease. Nat Rev Nephrol 6(2):96–106. doi:10.1038/nrneph.2009.214

Barua M, Cil O, Paterson AD, Wang K, He N, Dicks E, Parfrey P, Pei Y (2009) Family history of renal disease severity predicts the mutated gene in ADPKD. J Am Soc Nephrol JASN 20(8):1833–1838. doi:10.1681/ASN.2009020162

Ben-Dov IZ, Tan YC, Morozov P, Wilson PD, Rennert H, Blumenfeld JD, Tuschl T (2014) Urine microRNA as potential biomarkers of autosomal dominant polycystic kidney disease progression: description of miRNA profiles at baseline. PLoS One 9(1):e86856. doi:10.1371/journal.pone.0086856

Bergmann C, von Bothmer J, Ortiz Bruchle N, Venghaus A, Frank V, Fehrenbach H, Hampel T, Pape L, Buske A, Jonsson J, Sarioglu N, Santos A, Ferreira JC, Becker JU, Cremer R, Hoefele J, Benz MR, Weber LT, Buettner R, Zerres K (2011) Mutations in multiple PKD genes may explain early and severe polycystic kidney disease. J Am Soc Nephrol JASN 22(11):2047–2056. doi:10.1681/ASN.2010101080

Boertien WE, Meijer E, Zittema D, van Dijk MA, Rabelink TJ, Breuning MH, Struck J, Bakker SJ, Peters DJ, de Jong PE, Gansevoort RT (2012) Copeptin, a surrogate marker for vasopressin, is associated with kidney function decline in subjects with autosomal dominant polycystic kidney disease. Nephrol Dial Transplant 27(11):4131–4137. doi:10.1093/ndt/gfs070

Boertien WE, Meijer E, Li J, Bost JE, Struck J, Flessner MF, Gansevoort RT, Torres VE, Consortium for Radiologic Imaging Studies of Polycystic Kidney Disease C (2013) Relationship of copeptin, a surrogate marker for arginine vasopressin, with change in total kidney volume and GFR decline in autosomal dominant polycystic kidney disease: results from the CRISP cohort. Am J Kidney Dis Off J Nat Kidney Found 61(3):420–429. doi:10.1053/j.ajkd.2012.08.038

Chapman AB, Bost JE, Torres VE, Guay-Woodford L, Bae KT, Landsittel D, Li J, King BF, Martin D, Wetzel LH, Lockhart ME, Harris PC, Moxey-Mims M, Flessner M, Bennett WM, Grantham JJ (2012) Kidney volume and functional outcomes in autosomal dominant polycystic kidney disease. Clin J Am Soc Nephrol 7(3):479–486. doi:10.2215/CJN.09500911

Chapman AB, Devuyst O, Eckardt KU, Gansevoort RT, Harris T, Horie S, Kasiske BL, Odland D, Pei Y, Perrone RD, Pirson Y, Schrier RW, Torra R, Torres VE, Watnick T, Wheeler DC, Conference P (2015) Autosomal-dominant polycystic kidney disease (ADPKD): executive summary from a Kidney Disease: Improving Global Outcomes (KDIGO) controversies conference. Kidney Int 88(1):17–27. doi:10.1038/ki.2015.59

Cornec-Le Gall E, Audrezet MP, Chen JM, Hourmant M, Morin MP, Perrichot R, Charasse C, Whebe B, Renaudineau E, Jousset P, Guillodo MP, Grall-Jezequel A, Saliou P, Ferec C, Le Meur Y (2013) Type of PKD1 mutation influences renal outcome in ADPKD. J Am Soc Nephrol JASN 24(6):1006–1013. doi:10.1681/ASN.2012070650

Fick-Brosnahan GM, Tran ZV, Johnson AM, Strain JD, Gabow PA (2001) Progression of autosomal-dominant polycystic kidney disease in children. Kidney Int 59(5):1654–1662. doi:10.1046/j.1523-1755.2001.0590051654.x

Gabow PA, Duley I, Johnson AM (1992a) Clinical profiles of gross hematuria in autosomal dominant polycystic kidney disease. Am J Kidney Dis Off J Nat Kidney Found 20(2):140–143

Gabow PA, Johnson AM, Kaehny WD, Kimberling WJ, Lezotte DC, Duley IT, Jones RH (1992b) Factors affecting the progression of renal disease in autosomal-dominant polycystic kidney disease. Kidney Int 41(5):1311–1319

Gansevoort RT, Arici M, Benzing T, Birn H, Capasso G, Covic A, Devuyst O, Drechsler C, Eckardt KU, Emma F, Knebelmann B, Le Meur Y, Massy ZA, Ong AC, Ortiz A, Schaefer F, Torra R, Vanholder R, Wiecek A, Zoccali C, Van Biesen W (2016) Recommendations for the use of tolvaptan in autosomal dominant polycystic kidney disease: a position statement on behalf of the ERA-EDTA working groups on inherited kidney disorders and European renal best practice. Nephrol Dial Transplant 31(3):337–348. doi:10.1093/ndt/gfv456

Geberth S, Stier E, Zeier M, Mayer G, Rambausek M, Ritz E (1995) More adverse renal prognosis of autosomal dominant polycystic kidney disease in families with primary hypertension. J Am Soc Nephrol JASN 6(6):1643–1648

Harris PC, Bae KT, Rossetti S, Torres VE, Grantham JJ, Chapman AB, Guay-Woodford LM, King BF, Wetzel LH, Baumgarten DA, Kenney PJ, Consugar M, Klahr S, Bennett WM, Meyers CM, Zhang QJ, Thompson PA, Zhu F, Miller JP (2006) Cyst number but not the rate of cystic growth is associated with the mutated gene in autosomal dominant polycystic kidney disease. J Am Soc Nephrol JASN 17(11):3013–3019. doi:10.1681/ASN.2006080835

Hateboer N, v Dijk MA, Bogdanova N, Coto E, Saggar-Malik AK, San Millan JL, Torra R, Breuning M, Ravine D (1999) Comparison of phenotypes of polycystic kidney disease types 1 and 2. European PKD1-PKD2 Study Group. Lancet 353(9147):103–107

Helal I, McFann K, Reed B, Yan XD, Schrier RW, Fick-Brosnahan GM (2013) Serum uric acid, kidney volume and progression in autosomal-dominant polycystic kidney disease. Nephrol Dial Transplant 28(2):380–385. doi:10.1093/ndt/gfs417

Huang S, Chen L, Lu L, Li L (2016) The apelin-APJ axis: a novel potential therapeutic target for organ fibrosis. Clin Chim Acta 456:81–88. doi:10.1016/j.cca.2016.02.025

Hwang JH, Park HC, Jeong JC, Ha Baek S, Han MY, Bang K, Cho JY, Yu SH, Yang J, Oh KH, Hwang YH, Ahn C (2013) Chronic asymptomatic pyuria precedes overt urinary tract infection and deterioration of renal function in autosomal dominant polycystic kidney disease. BMC Nephrol 14:1. doi:10.1186/1471-2369-14-1

Hwang YH, Conklin J, Chan W, Roslin NM, Liu J, He N, Wang K, Sundsbak JL, Heyer CM, Haider M, Paterson AD, Harris PC, Pei Y (2015) Refining genotype-phenotype correlation in autosomal dominant polycystic kidney disease. J Am Soc Nephrol JASN. doi:10.1681/ASN.2015060648

Idrizi A, Barbullushi M, Petrela E, Kodra S, Koroshi A, Thereska N (2009) The influence of renal manifestations to the progression of autosomal dominant polycystic kidney disease. Hippokratia 13(3):161–164

Idrizi A, Barbullushi M, Koroshi A, Dibra M, Bolleku E, Bajrami V, Xhaferri X, Thereska N (2011) Urinary tract infections in polycystic kidney disease. Med Arh 65(4):213–215

Irazabal MV, Rangel LJ, Bergstralh EJ, Osborn SL, Harmon AJ, Sundsbak JL, Bae KT, Chapman AB, Grantham JJ, Mrug M, Hogan MC, El-Zoghby ZM, Harris PC, Erickson BJ, King BF, Torres VE, Investigators C (2015) Imaging classification of autosomal dominant polycystic kidney disease: a simple model for selecting patients for clinical trials. J Am Soc Nephrol JASN 26(1):160–172. doi:10.1681/ASN.2013101138

Johnson AM, Gabow PA (1997) Identification of patients with autosomal dominant polycystic kidney disease at highest risk for end-stage renal disease. J Am Soc Nephrol JASN 8(10):1560–1567

Kistler AD, Mischak H, Poster D, Dakna M, Wuthrich RP, Serra AL (2009a) Identification of a unique urinary biomarker profile in patients with autosomal dominant polycystic kidney disease. Kidney Int 76(1):89–96. doi:10.1038/ki.2009.93

Kistler AD, Poster D, Krauer F, Weishaupt D, Raina S, Senn O, Binet I, Spanaus K, Wuthrich RP, Serra AL (2009b) Increases in kidney volume in autosomal dominant polycystic kidney disease can be detected within 6 months. Kidney Int 75(2):235–241. doi:10.1038/ki.2008.558

Kistler AD, Serra AL, Siwy J, Poster D, Krauer F, Torres VE, Mrug M, Grantham JJ, Bae KT, Bost JE, Mullen W, Wuthrich RP, Mischak H, Chapman AB (2013) Urinary proteomic biomarkers for diagnosis and risk stratification of autosomal dominant polycystic kidney disease: a multicentric study. PLoS One 8(1):e53016. doi:10.1371/journal.pone.0053016

Klahr S, Breyer JA, Beck GJ, Dennis VW, Hartman JA, Roth D, Steinman TI, Wang SR, Yamamoto ME (1995) Dietary protein restriction, blood pressure control, and the progression of polycystic kidney disease. Modification of Diet in Renal Disease Study Group. J Am Soc Nephrol JASN 5(12):2037–2047

Kobori H, Alper AB Jr, Shenava R, Katsurada A, Saito T, Ohashi N, Urushihara M, Miyata K, Satou R, Hamm LL, Navar LG (2009) Urinary angiotensinogen as a novel biomarker of the intrarenal renin-angiotensin system status in hypertensive patients. Hypertension 53(2):344–350. doi:10.1161/HYPERTENSIONAHA.108.123802

Kocer D, Karakukcu C, Ozturk F, Eroglu E, Kocyigit I (2016) Evaluation of fibrosis markers: apelin and transforming growth factor-beta1 in autosomal dominant polycystic kidney disease patients. The Apher Dial Off Peer-reviewed J Int Soc Apher Jpn Soc Apher Jpn Soc Dial Ther. doi:10.1111/1744-9987.12412

Kocyigit I, Yilmaz MI, Orscelik O, Sipahioglu MH, Unal A, Eroglu E, Kalay N, Tokgoz B, Axelsson J, Oymak O (2013) Serum uric acid levels and endothelial dysfunction in patients with autosomal dominant polycystic kidney disease. Nephron Clin Pract 123(3–4):157–164. doi:10.1159/000353730

Lacquaniti A, Chirico V, Lupica R, Buemi A, Loddo S, Caccamo C, Salis P, Bertani T, Buemi M (2013) Apelin and copeptin: two opposite biomarkers associated with kidney function decline and cyst growth in autosomal dominant polycystic kidney disease. Peptides 49:1–8. doi:10.1016/j.peptides.2013.08.007

Lakhia R, Hajarnis S, Williams D, Aboudehen K, Yheskel M, Xing C, Hatley ME, Torres VE, Wallace DP, Patel V (2015) MicroRNA-21 aggravates cyst growth in a model of polycystic kidney disease. J Am Soc Nephrol JASN. doi:10.1681/ASN.2015060634

Li X (2011) Epigenetics and autosomal dominant polycystic kidney disease. Biochim Biophys Acta 1812(10):1213–1218. doi:10.1016/j.bbadis.2010.10.008

Liu M, Shi S, Senthilnathan S, Yu J, Wu E, Bergmann C, Zerres K, Bogdanova N, Coto E, Deltas C, Pierides A, Demetriou K, Devuyst O, Gitomer B, Laakso M, Lumiaho A, Lamnissou K, Magistroni R, Parfrey P, Breuning M, Peters DJ, Torra R, Winearls CG, Torres VE, Harris PC, Paterson AD, Pei Y (2010) Genetic variation of DKK3 may modify renal disease severity in ADPKD. J Am Soc Nephrol JASN 21(9):1510–1520. doi:10.1681/ASN.2010030237

Meijer E, Boertien WE, Nauta FL, Bakker SJ, van Oeveren W, Rook M, van der Jagt EJ, van Goor H, Peters DJ, Navis G, de Jong PE, Gansevoort RT (2010) Association of urinary biomarkers with disease severity in patients with autosomal dominant polycystic kidney disease: a cross-sectional analysis. Am J Kidney Dis Off J Nat Kidney Found 56(5):883–895. doi:10.1053/j.ajkd.2010.06.023

Meijer E, Bakker SJ, van der Jagt EJ, Navis G, de Jong PE, Struck J, Gansevoort RT (2011) Copeptin, a surrogate marker of vasopressin, is associated with disease severity in autosomal dominant polycystic kidney disease. Clin J Am Soc Nephrol 6(2):361–368. doi:10.2215/CJN.04560510

Nakajima A, Lu Y, Kawano H, Horie S, Muto S (2015) Association of arginine vasopressin surrogate marker urinary copeptin with severity of autosomal dominant polycystic kidney disease (ADPKD). Clin Exp Nephrol 19(6):1199–1205. doi:10.1007/s10157-015-1101-7

Norman J (2011) Fibrosis and progression of autosomal dominant polycystic kidney disease (ADPKD). Biochim Biophys Acta 1812(10):1327–1336. doi:10.1016/j.bbadis.2011.06.012

Orskov B, Borresen ML, Feldt-Rasmussen B, Ostergaard O, Laursen I, Strandgaard S (2010) Estimating glomerular filtration rate using the new CKD-EPI equation and other equations in patients with autosomal dominant polycystic kidney disease. Am J Nephrol 31(1):53–57. doi:10.1159/000256657

Ozkok A, Akpinar TS, Tufan F, Kanitez NA, Uysal M, Guzel M, Caliskan Y, Alisir S, Yazici H, Ecder T (2013) Clinical characteristics and predictors of progression of chronic kidney disease in autosomal dominant polycystic kidney disease: a single center experience. Clin Exp Nephrol 17(3):345–351. doi:10.1007/s10157-012-0706-3

Parikh CR, Dahl NK, Chapman AB, Bost JE, Edelstein CL, Comer DM, Zeltner R, Tian X, Grantham JJ, Somlo S (2012) Evaluation of urine biomarkers of kidney injury in polycystic kidney disease. Kidney Int 81(8):784–790. doi:10.1038/ki.2011.465

Park EY, Woo YM, Park JH (2011) Polycystic kidney disease and therapeutic approaches. BMB Rep 44(6):359–368. doi:10.5483/BMBRep.2011.44.6.359

Park HC, Hwang JH, Kang AY, Ro H, Kim MG, An JN, In Park J, Kim SH, Yang J, Oh YK, Oh KH, Noh JW, Cheong HI, Hwang YH, Ahn C (2012) Urinary N-acetyl-beta-D glucosaminidase as a surrogate marker for renal function in autosomal dominant polycystic kidney disease: 1 year prospective cohort study. BMC Nephrol 13:93. doi:10.1186/1471-2369-13-93

Park HC, Kang AY, Jang JY, Kim H, Han M, Oh KH, Kim SH, Noh JW, Cheong HI, Hwang YH, Ahn C (2015) Increased urinary Angiotensinogen/Creatinine (AGT/Cr) ratio may be associated with reduced renal function in autosomal dominant polycystic kidney disease patients. BMC Nephrol 16:86. doi:10.1186/s12882-015-0075-8

Paterson AD, Pei Y (1998) Is there a third gene for autosomal dominant polycystic kidney disease? Kidney Int 54(5):1759–1761. doi:10.1046/j.1523-1755.1998.00166.x

Paul BM, Consugar MB, Ryan Lee M, Sundsbak JL, Heyer CM, Rossetti S, Kubly VJ, Hopp K, Torres VE, Coto E, Clementi M, Bogdanova N, de Almeida E, Bichet DG, Harris PC (2014) Evidence of a third ADPKD locus is not supported by re-analysis of designated PKD3 families. Kidney Int 85(2):383–392. doi:10.1038/ki.2013.227

Pei Y, Lan Z, Wang K, Garcia-Gonzalez M, He N, Dicks E, Parfrey P, Germino G, Watnick T (2012) A missense mutation in PKD1 attenuates the severity of renal disease. Kidney Int 81(4):412–417. doi:10.1038/ki.2011.370

Perez-Oller L, Torra R, Badenas C, Mila M, Darnell A (1999) Influence of the ACE gene polymorphism in the progression of renal failure in autosomal dominant polycystic kidney disease. Am J Kidney Dis Off J Nat Kidney Found 34(2):273–278. doi:10.1053/AJKD03400273

Persu A, Stoenoiu MS, Messiaen T, Davila S, Robino C, El-Khattabi O, Mourad M, Horie S, Feron O, Balligand JL, Wattiez R, Pirson Y, Chauveau D, Lens XM, Devuyst O (2002) Modifier effect of ENOS in autosomal dominant polycystic kidney disease. Hum Mol Genet 11(3):229–241

Romaker D, Puetz M, Teschner S, Donauer J, Geyer M, Gerke P, Rumberger B, Dworniczak B, Pennekamp P, Buchholz B, Neumann HP, Kumar R, Gloy J, Eckardt KU, Walz G (2009) Increased expression of secreted frizzled-related protein 4 in polycystic kidneys. J Am Soc Nephrol JASN 20(1):48–56. doi:10.1681/ASN.2008040345

Rossetti S, Kubly VJ, Consugar MB, Hopp K, Roy S, Horsley SW, Chauveau D, Rees L, Barratt TM, van't Hoff WG, Niaudet P, Torres VE, Harris PC (2009) Incompletely penetrant PKD1 alleles suggest a role for gene dosage in cyst initiation in polycystic kidney disease. Kidney Int 75(8):848–855. doi:10.1038/ki.2008.686

Ruggenenti P, Gaspari F, Cannata A, Carrara F, Cella C, Ferrari S, Stucchi N, Prandini S, Ene-Iordache B, Diadei O, Perico N, Ondei P, Pisani A, Buongiorno E, Messa P, Dugo M, Remuzzi G, Group G-AS (2012) Measuring and estimating GFR and treatment effect in ADPKD patients: results and implications of a longitudinal cohort study. PLoS One 7(2):e32533. doi:10.1371/journal.pone.0032533

Rule AD, Torres VE, Chapman AB, Grantham JJ, Guay-Woodford LM, Bae KT, Klahr S, Bennett WM, Meyers CM, Thompson PA, Miller JP, Consortium C (2006) Comparison of methods for determining renal function decline in early autosomal dominant polycystic kidney disease: the consortium of radiologic imaging studies of polycystic kidney disease cohort. J Am Soc Nephrol JASN 17(3):854–862. doi:10.1681/ASN.2005070697

Salih M, Demmers JA, Bezstarosti K, Leonhard WN, Losekoot M, van Kooten C, Gansevoort RT, Peters DJ, Zietse R, Hoorn EJ, Consortium D (2016) Proteomics of urinary vesicles links plakins and complement to polycystic kidney disease. J Am Soc Nephrol JASN. doi:10.1681/ASN.2015090994

Schrier RW (2011) Hypertension and autosomal dominant polycystic kidney disease. Am J Kidney Dis Off J Nat Kidney Found 57(6):811–813. doi:10.1053/j.ajkd.2011.02.379

Schrier RW, Johnson AM, McFann K, Chapman AB (2003) The role of parental hypertension in the frequency and age of diagnosis of hypertension in offspring with autosomal-dominant polycystic kidney disease. Kidney Int 64(5):1792–1799. doi:10.1046/j.1523-1755.2003.00264.x

Schrier RW, Abebe KZ, Perrone RD, Torres VE, Braun WE, Steinman TI, Winklhofer FT, Brosnahan G, Czarnecki PG, Hogan MC, Miskulin DC, Rahbari-Oskoui FF, Grantham JJ, Harris PC, Flessner MF, Bae KT, Moore CG, Chapman AB, Investigators H-PT (2014) Blood pressure in early autosomal dominant polycystic kidney disease. N Engl J Med 371(24):2255–2266. doi:10.1056/NEJMoa1402685

Shamshirsaz AA, Reza Bekheirnia M, Kamgar M, Johnson AM, McFann K, Cadnapaphornchai M, Nobakhthaghighi N, Schrier RW (2005) Autosomal-dominant polycystic kidney disease in infancy and childhood: progression and outcome. Kidney Int 68(5):2218–2224. doi:10.1111/j.1523-1755.2005.00678.x

Spithoven EM, van Gastel MD, Messchendorp AL, Casteleijn NF, Drenth JP, Gaillard CA, de Fijter JW, Meijer E, Peters DJ, Kappert P, Renken RJ, Visser FW, Wetzels JF, Zietse R, Gansevoort RT, Consortium D (2015) Estimation of total kidney volume in autosomal dominant polycystic kidney disease. Am J Kidney Dis Off J Nat Kidney Found 66(5):792–801. doi:10.1053/j.ajkd.2015.06.017

Spithoven EM, Meijer E, Boertien WE, Sinkeler SJ, Tent H, de Jong PE, Navis G, Gansevoort RT (2013) Tubular secretion of creatinine in autosomal dominant polycystic kidney disease: consequences for cross-sectional and longitudinal performance of kidney function estimating equations. Am J Kidney Dis Off J Nat Kidney Found 62(3):531–540. doi:10.1053/j.ajkd.2013.03.030

Sun L, Zhu J, Wu M, Sun H, Zhou C, Fu L, Xu C, Mei C (2015) Inhibition of MiR-199a-5p reduced cell proliferation in autosomal dominant polycystic kidney disease through targeting CDKN1C. Med Sci Mon Int Med J Exp Clin Res 21:195–200. doi:10.12659/MSM.892141

Torres VE, Harris PC, Pirson Y (2007) Autosomal dominant polycystic kidney disease. Lancet 369(9569):1287–1301. doi:10.1016/S0140-6736(07)60601-1

Tracz MJ, Alam J, Nath KA (2007) Physiology and pathophysiology of heme: implications for kidney disease. J Am Soc Nephrol JASN 18(2):414–420. doi:10.1681/ASN.2006080894

Wang D, Strandgaard S, Borresen ML, Luo Z, Connors SG, Yan Q, Wilcox CS (2008) Asymmetric dimethylarginine and lipid peroxidation products in early autosomal dominant polycystic kidney disease. Am J Kidney Dis Off J Nat Kidney Found 51(2):184–191. doi:10.1053/j.ajkd.2007.09.020

Woo YM, Park JH (2013) microRNA biomarkers in cystic diseases. BMB Rep 46(7):338–345

Woo YM, Bae JB, Oh YH, Lee YG, Lee MJ, Park EY, Choi JK, Lee S, Shin Y, Lyu J, Jung HY, Lee YS, Hwang YH, Kim YJ, Park JH (2014) Genome-wide methylation profiling of ADPKD identified epigenetically regulated genes associated with renal cyst development. Hum Genet 133(3):281–297. doi:10.1007/s00439-013-1378-0

Zheng D, Wolfe M, Cowley BD Jr, Wallace DP, Yamaguchi T, Grantham JJ (2003) Urinary excretion of monocyte chemoattractant protein-1 in autosomal dominant polycystic kidney disease. J Am Soc Nephrol JASN 14(10):2588–2595

Zschiedrich S, Budde K, Nurnberger J, Wanner C, Sommerer C, Kunzendorf U, Banas B, Hoerl WH, Obermuller N, Arns W, Pavenstadt H, Gaedeke J, Lindner TH, Faerber L, Wimmer P, Stork R, Eckardt KU, Walz G (2016) Secreted frizzled-related protein 4 predicts progression of autosomal dominant polycystic kidney disease. Nephrol Dial Transplant 31(2):284–289. doi:10.1093/ndt/gfv077

Chapter 9
Clinical Trials and a View Toward the Future of ADPKD

Hyunsuk Kim and Young-Hwan Hwang

Abstract In light of the advances in the understanding of cystogenesis in clinical syndromes, potential therapeutic targets have been proposed. Among ciliopathies, autosomal dominant polycystic kidney disease (ADPKD) is the most common hereditary disease, and is characterized by the progressive enlargement of bilateral renal cysts, resulting in end-stage kidney failure. Progress in genetics and molecular pathobiology has enabled the development of therapeutic agents that can modulate aberrant molecular pathways. Recently, clinical trials using somatostatin analogs and vasopressin receptor antagonists were conducted, and resulted in the approval of tolvaptan in managing kidney disease in some countries. We will summarize the developments of therapeutic agents based on pathogenesis, and discuss recent findings in clinical trials. Moreover, issues such as the timing of the intervention and outcome assessment will be discussed.

Keywords Polycystic kidney • Autosomal dominant • Cystogenesis • Therapeutics • Clinical trials

Autosomal dominant polycystic kidney disease (ADPKD) is the most common hereditary kidney disease worldwide. Multiple cysts develop and grow in both kidneys, resulting in the deterioration of kidney function and renal failure. The incidence varies in different countries, but it develops at a rate of roughly one in 1000 live births (Bennett 2009). Kidney failure develops in approximately half of ADPKD patients by the age of 60, and requires renal replacement therapy. In South Korea, ADPKD is responsible for approximately 2 % of prevalent end-stage kidney diseases, following only diabetes, hypertension, and glomerulonephritis in the proportion of end-stage kidney disease that it causes. Causative mutations have been found

H. Kim (✉)
Department of Internal Medicine, Seoul National University Hospital,
101, Daehak-ro, Jongno-gu, Seoul 03080, South Korea
e-mail: keeee@hanmail.net

Y.-H. Hwang
Department of Internal Medicine, Eulji General Hospital,
14, Hangeulbiseok-gil, Nowon-gu, Seoul 01830, South Korea
e-mail: nephro.hwang@gmail.com

J.H. Park, C. Ahn (eds.), *Cystogenesis*, Advances in Experimental Medicine and Biology 933, DOI 10.1007/978-981-10-2041-4_9

in the PKD1 and PKD2 genes that encode polycystin-1 and polycystin-2 proteins, respectively. Once disease genes and their coding proteins were identified, aberrant changes in various intracellular signaling pathways due to gene mutation were reported. In addition, a wide array of experimental model systems has been developed and used in the search for therapeutic targets, which has been fully described in previous chapters.

In terms of management of ADPKD, no pathogenesis-based approach was available until recently. Most current recommendations involve the control of blood pressure and general lifestyle modification, including water intake. Since the publication of promising results from the TEMPO 3/4 trial (Torres et al. 2011; 2012), disease-specific therapy is expected to be implemented in the near future. However, the TEMPO trial raised many issues surrounding the clinical management of ADPKD, such as the long-term risks and benefits of candidate drugs, the timing of treatment, outcome evaluation, and selection of high-risk patients.

In this review, the authors will focus on recent clinical trials in patients with ADPKD, and discuss several issues related with clinical management.

9.1 Conventional Management

9.1.1 Water

Arginine vasopressin (AVP), or antidiuretic hormone (ADH), binds to the specific AVP receptors (V2R) at the renal collecting duct to increase cellular cyclic AMP (cAMP). An increased level of cAMP enhances protein kinase A (PKA) activity and subsequent signaling pathways, leading to cyst growth. Thus, theoretically, high water intake has been thought to inhibit vasopressin secretion and, in turn, cyst growth in the kidneys. The benefit of high water intake was first proposed in a PCK rat model (Nagao et al. 2006). PCK rats in a high water group, in which 3.5-fold increased water intake was observed, showed reduced kidney size and improved kidney function. This result was similar to the effects of AVP receptor blockers (Torres 2005) In order to achieve all-day-long suppression of AVP and urine osmolality as low as 250 mOsm/kg, water intake was encouraged to produce as much as 2.5–4 L of urine a day (Torres et al. 2009).

However, it is still unclear whether high water intake is beneficial to slow renal progression in ADPKD patients. In a recent one-year follow-up study, 18 ADPKD patients were randomly assigned to a high water intake group (50 mL/kg or more a day, or 2.5–3.0 L a day), while 16 patients were allocated to the free water intake group. Compared to the free water intake group, the high water intake group showed a significantly reduced copeptin level, which represented the blood AVP level. However, kidney volume, which was a primary outcome variable, and kidney function, which was a secondary endpoint, failed to show significant differences between the two groups. In fact, the rate of change in total kidney volume (TKV) and estimated glomerular filtration rate (eGFR) was accelerated in the high water intake group during the study period, compared to the pre-study period (Higashihara et al. 2014). This may be explained by the possibility that the water intake alone is insufficient to suppress AVP adequately (Torres et al. 2011). In parallel with this finding, a water intake of 3 L a day for one week (chronic water loading) was shown to lower urine osmolality to a mean of 270 mOsm/kg, but the changes in the 24-h urine cAMP excretion, which reflected the blood vasopressin activity level, was insignificant (Barash et al. 2010). The effects of increased water intake on the renal progression in ADPKD should be clarified through large-scale clinical trials. A more detailed review of water handling in ADPKD can be found in Torres et al. (2009).

9.1.2 Blood Pressure Control

Hypertension is the most common renal manifestation of ADPKD, observed in 65–93 % of patients, and is a major risk factor for renal failure and cardiovascular complications. Blood pressure increases even in relatively young patients, as well as the activation of the renal renin-angiotensin system, were demonstrated at the early stage of disease (Chapman et al. 1990). Therefore, strict control of blood pressure has been the mainstay of clinical management in ADPKD.

In a 7-year follow-up study, in which 75 ADPKD patients were enrolled, the group whose blood pressure was controlled at a target of 120/80 mm Hg or below showed a significant decrease in left ventricular hypertrophy (Schrier et al. 2002). The Modification of Diet in Renal Disease (MDRD) extended follow-up study evaluated the effect of a lower target blood pressure (mean arterial pressure <92 vs. <107 mm Hg) on the progression to kidney failure in patients with GFR of 13–55 mL/min/1.73 m^2. When patients with ADPKD (n=200) were analyzed, a lower target blood pressure was associated with significant risk reduction (Sarnak et al. 2005). Based on these results, intensive blood pressure control to a target of 130/80 mm Hg has been proposed, preferably using renin-angiotensin system (RAS) blockers – angiotensin converting enzyme inhibitors (ACEIs) or angiotensin receptor blockers (ARBs). However, the above studies were under-powered or not designed for evaluation of ADPKD; thus no evidence was available to support the recommendations. Moreover, aggressive blood pressure control to < 120/80 mm Hg may increase cardiovascular risk in some patients (Yusuf et al. 2008; Redon et al. 2012).

To address this issue, the HALT-PKD study (NCT 00283686 & NCT 01885559) was launched in 2006, and the results of this large-scale trial were published in 2014. This study, sponsored by the U.S. National Institutes of Health, enrolled ADPKD patients with hypertension, and followed up for five years (Schrier et al. 2014a; Torres et al. 2014; Chapman et al. 2010b). The subjects were assigned to either Study A or B, based on their baseline kidney function (Fig. 9.1). In Study A (under 50 years of age, baseline eGFR > 60 mL/min/1.73 m^2), the subjects were randomly assigned to one of four groups (2 × 2 factorial design): blood pressure (BP)

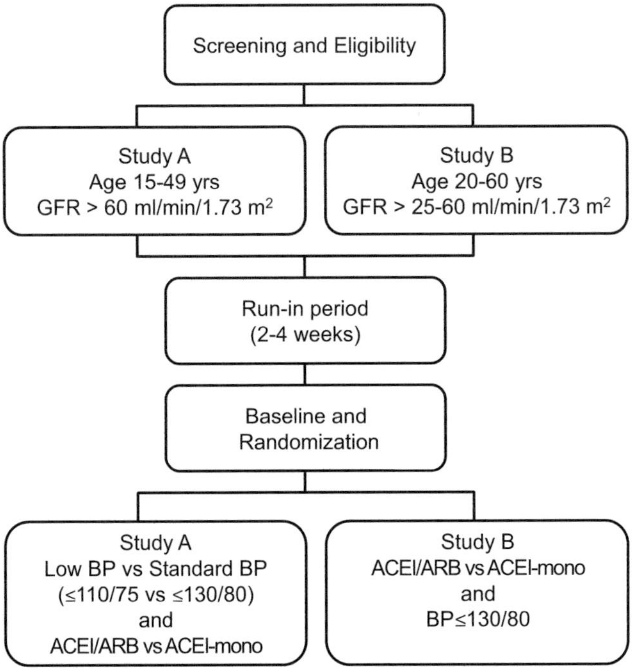

Fig. 9.1 Design of HALT-PKD Study A and B (Chapman et al. 2010b)

goals of 110/75 (low BP group) versus 130/80 mm Hg (standard BP group), and ACEI/ARB-combination vs ACEI-monotherapy. In Study B (18–64 years of age, eGFR of 25–60 mL/min/1.73 m^2), blood pressure was controlled at 130/80 mm Hg or below, and ACEI/ARB-combination therapy was compared with ACEI-monotherapy. In Study A, the primary endpoint was percentage change in TKV measured by MRI; and in Study B, the composite of time to either 50 % reduction of baseline eGFR, end-stage renal failure, or death.

A total of 558 patients with early-stage ADPKD participated in HALT Study A. Both the standard blood pressure group (120/70 – 130/80 mm Hg) and the low blood pressure group (95/60 – 110/75 mm Hg) adequately maintained their blood pressure goals. The percent change in TKV was significantly lower in the low blood pressure group than the standard group (5.6 % vs. 6.6 % per year, $P=0.006$). In addition, the left ventricular mass index (LVMI) fell (−1.17 vs. −0.57 g/m^2 per year, $P<0.001$), and the albuminuria level also fell significantly (−3.77 % vs. 2.43 %, $P<0.001$) in the low BP group. However, the eGFR fell by approximately 3 mL/min/1.73 m^2 annually in the two groups, showing no significant difference. In contrast, no outcome variables were shown to be different between the ACEI/ARB-combination and ACEI monotherapy groups. In HALT Study B, 486 patients at the advanced chronic kidney disease (CKD) stage were randomly assigned to either combination or monotherapy group. Monotherapy with an ACEI was comparable to the ACEI/ARB combination regimen in terms of BP control and composite primary outcome. Urinary aldosterone excretion was similarly lowered by both regimens. Adverse events, including hyperkalemia and acute kidney injury, were reported similarly in the two groups.

In summary, the rigorous control of blood pressure slightly but significantly reduced the rate of increase in TKV in patients with early ADPKD and may reduce the development of cardiovascular complications. Despite the fact that the ACEI/ARB-combination therapy failed to show additional favorable effects, it was found to be safe even in patients with CKD stage 3.

For now, it would be prudent to target a blood pressure of 130/80 mm Hg or below, and out-of-office blood pressure recording should be encouraged for monitoring, as in the HALT-PKD study. RAS blockers are recommended as first-line agents, and second-line drugs can be chosen depending on comorbid diseases and the individual profile. It is of note that diuretics can activate the renal renin-angiotensin system, increase AVP, and activate cAMP (Ecder et al. 2001). Non-dihydropyridine calcium channel blockers, such as verapamil, may accelerate cyst growth, and they should thus be avoided (Nagao et al. 2008).

The HALT-PKD study clearly has shown that a conventional approach alone cannot prevent renal progression in ADPKD, and the development of disease-specific anti-proliferative agents is needed. 9.2 Disease-specific Interventions

9.2 Recent Clinical Trials

9.2.1 mTOR Inhibitors

The mammalian target of rapamycin (mTOR) is known to regulate various signaling mechanisms associated with cell proliferation. Thus mTOR inhibitors were one of the attractive candidates for drug development for ADPKD. Rapamycin (or sirolimus) was shown to inhibit cyst epithelial cell proliferation, to reduce cyst growth, and to prevent the deterioration of kidney function in preclinical studies and pilot clinical studies (Tao et al. 2005; Wahl et al. 2006). However, two large-scale randomized clinical trials failed to confirm the efficacy of mTOR inhibitors. In the SUISSE study, Serra et al. carefully enrolled 100 subjects aged 18–40 years with a creatinine clearance > 70 mL/min, but with high risk of progression, as verified by a TKV increase of more than 2 % during a 6-month run-in period. The 18-month administration of sirolimus resulted in no significant differences in kidney volume or function (Serra et al. 2010). Another two-year randomized controlled trial using everolimus recruited ADPKD patients in the advanced CKD stage with a mean eGFR of 53–56 mL/min/1.73 m^2 and a mean TKV of 1900–2000 mL. In the everolimus group, the rate of kidney volume change was significantly lower in the first year ($P = 0.02$), but no difference from the control was observed at the end of the study. Contrary to the hypothesis, the everolimus group showed a faster decline of the eGFR (−5.5 vs. −3.5 mL/min/1.73 m^2 per year, $P < 0.001$) (Walz et al. 2010).

Practically, mTOR inhibitors were associated with various adverse effects, including anemia (relative ratio [RR] 3.41), angioedema (RR 13.39), diarrhea (RR 1.70), hyperlipidemia (RR 5.68), and oral ulcers (RR 6.77), compared with placebo (Bolignano et al. 2015). Poor adherence in the everolimus trial may be explained by the high rate of these side effects. Sirolimus was well tolerated in the SUISSE trial, but study drug blood levels were at the lower end of the target range of 4–10 ng/mL (Watnick and Germino 2010). In contrast, even the usual dose of sirolimus was not able to inhibit the mTOR pathway in the kidneys (Canaud et al. 2010). It is of note that the effective dose in animal experiments was equivalent to 10 times the standard dose for humans (Novalic et al. 2012). To maximize the effects of cyst cell inhibition with reduced side effects, a delivery method involving folate-conjugated rapamycin was advocated but also failed (Shillingford et al. 2012).

9.2.2 Vasopressin Receptor-2 Antagonists (V2RAs)

V2R is mainly localized at the basolateral membrane of the principal cells of the collecting ducts in the kidneys. Activation of adenylyl cyclases leads to accumulation of cellular cAMP and water reabsorption through aquaporin water channels. V2RAs, or vaptans, can block the interaction with AVP, leading to aquaresis. V2RAs were developed as aquaretics for dilutional hyponatremia, such as the syndrome of inappropriate secretion of antidiuretic hormone or congestive heart failure (Berl 2015; Schrier et al. 2006) Researchers have been interested in the defects in urine concentration in patients with ADPKD and their elevated levels of blood AVP and cellular cAMP. In various preclinical studies, V2RAs were demonstrated to inhibit cell proliferation, fluid secretion into cysts, and cyst growth. In addition, when AVP-knockout rats were cross-bred with PCK rats, cyst growth was significantly reduced in PCK AVP (-/-) double knockout rats (Wang et al. 2008). Based on the proven central role of AVP in PKD, the TEMPO 3/4 study for evaluating the efficacy and safety of tolvaptan was launched in 2007, and the results were recently published (Torres et al. 2011; 2016; Muto et al. 2015). A total of 1445 ADPKD patients who were aged 50 or younger with TKV > 750 mL and a Cockcroft-Gault eGFR > 60 mL/min participated in the study. After 3 years, kidney volume increased by 5.5 % annually in the placebo group, and by 2.8 % in the treatment group, resulting in an approximately 50 % reduction in TKV change. For the renal function outcome, worsening kidney function defined by a 25 % reduction in the reciprocal of the serum creatinine (sCr) at the end of the dose-titration period, and the change in the slope of 1/sCr, were used as endpoints. Tolvaptan reduced the event rate of worsening kidney function (2 events per 100 person-years vs. 5 in the placebo group; hazard ratio [HR] 0.39) and slowed the rate of eGFR decline (treatment effect, ~1.0 mL/min/1.73 m^2) over 3 years. The results were similar in a sub-analysis of the Japanese population (n = 177) (Muto et al. 2015). Tolvaptan reduced annual TKV growth by

1.99 %, 3.12 %, and 2.61 % per year (all $P<0.001$) and eGFR decline by 0.40 in CKD1 ($P=0.23$), 1.13 in CKD2 ($P<0.001$) and 1.66 mL/min per 1.73 m^2 per year in CKD3 ($P<0.001$), with a trend toward a positive subgroup-treatment interaction ($P=0.07$) across CKD1, CKD2, and CKD3 (Torres et al. 2016). However, aquaretic side effects, including polyuria, were more frequently observed in the tolvaptan group, and accordingly, the drop-out rate reached 23 % (placebo group: 13.8 %). This higher drop-out rate in the tolvaptan group made it difficult to interpret the data; in other words, missing data from drop-out may influence the estimation of the risks. Moreover, unexpected liver dysfunction was found during the study, which had not been reported in previous tolvaptan trials. Overall, three cases of drug-induced hepatotoxicity were confirmed at 3–14 months (Watkins et al. 2015). Despite the lack of observed progression to severe liver failure, the U.S. FDA predicted the risk of severe liver failure at a rate of one in 3000 users, and recommended that appropriate risk evaluation and mitigation strategies (REMS) be performed (Gansevoort et al. 2016). The TEMPO-extension study is currently underway in the U.S.

9.2.3 Somatostatin Analogues (SAs)

Somatostatin inhibits cellular cAMP via mainly sst2 in kidney tissues, the bile duct, and liver cyst cells. In a few preclinical and pilot clinical studies, an SA was demonstrated to suppress kidney and liver cyst growth in PKD (Ruggenenti et al. 2005). Italian researchers conducted a six-month, crossover-designed study to test the effect on kidney volume change and tolerability of octreotide (Ruggenenti et al. 2005). Twelve subjects had advanced PKD with a TKV of ~2.4 L and mean sCr of 1.9 mg/dL. Promising efficacy and safety outcomes prompted them to conduct the ALADIN trial (NCT00309283), which was a three-year, placebo-controlled study (Caroli et al. 2013). Five large university hospitals in Italy participated in the study, and adults with a MDRD eGFR>40 mL/min/1.73 m^2 were randomly assigned to 40 mg of octreotide monthly ($n=40$) or placebo ($n=39$). At one year follow-up, changes in TKV and total cyst volume (TCV) in the octreotide group were significantly less than those in the control group (46.2 vs. 43.7 mL, $P=0.032$). Whereas, at the 3-year follow-up, there was no significant difference. When the data were further probed according to the study phase, the effect on TKV changes was most prominent in the first year of intervention. This finding corresponds to the observation in the TEMPO study; inhibition of fluid secretion seems be responsible for the decrease in TKV in the early phase. Interestingly, the trajectory of GFR change in the octreotide group was not linear, in contrast to the linear pattern in the control group; the slope of the GFR change was faster in the octreotide group during the first year, and then it stabilized. With regard to the correlation between one-year and one-to-three-year GFR changes, the greater one-year GFR reduction resulted in a lesser one-to-three-year GFR slope in the octreotide group. This finding is analogous to the bi-phasic response to the ACEI treatment in diabetic nephropathy (Nielsen et al. 1997). An extended study (NCT01377246) with 98 subjects has already begun, and other clinical trials, DIPAK1 (NCT01616927) and LIPS (NCT02127437), using lanreotide are currently underway in Europe (Meijer et al. 2014).

SAs have a more favorable profile in terms of side effects compared to other candidate drugs, such as mTOR inhibitors and tolvaptan. Moreover, the somatostatin receptors are present in the hepatobiliary ducts; therefore SAs can inhibit liver cyst growth, in contrast to tolvaptan (van Keimpema et al. 2009; Gevers et al. 2012).

9.2.4 New Hopes

Other than the aforementioned drugs, KD019, triptolide, curcumin, and spironolactone are currently undergoing clinical trials, and several candidate drugs are in preclinical trials (Chang and Ong 2013; Irazabal and Torres 2013) (Table 9.1). The Src kinase inhibitor (bosutinib) is a novel tyrosine kinase inhibitor that inhibited cyst growth in PKD mouse models (Sweeney et al. 2008; Elliott et al. 2011). This drug has been tested in breast and pancreatic cancer, as well as hematologic malignancy. A phase II, multicenter, randomized, double-blind, placebo-controlled clinical trial has recently ended and is expected to be published soon (NCT01233869). Tesevatinib (KD019), another novel tyrosine kinase inhibitor is in its phase 1/2 trial (NCT01559363).

Triptolide is an extract of Tripterygium wilfordii and has been used widely as an effective antiproteinuric in China. Triptolide reduces cyst formation in a neonatal-to-adult transition Pkd1 model of ADPKD (Leuenroth et al. 2010). In a pilot study, triptolide decreased proteinuria in ADPKD patients without a significant effect on TKV or eGFR (Chen et al. 2014). Recently, a randomized controlled trial on the effect of triptolide was conducted but was terminated due to the high dropout rate (NCT00801268) and another trial is under way (NCT02115659).

Curcumin is a polyphenol natural product, which is known to modulate several pathways (mTOR, WNT, STAT3) altered in ADPKD. Its effect was shown in some in vitro and in vivo experiments (Leonhard et al. 2011; Gao et al. 2011) and is now on clinical trial (NCT02494141).

Spironolactone, a diuretic agent, is on clinical trial to determine its effectiveness in blocking aldosterone for improving the health and function of arteries and renal progression in ADPKD (NCT01853553).

It is also of interest whether the combined use of drugs from different classes will show higher efficacy or synergism (Aguiari et al. 2013; Hopp et al. 2015; Chrispijn and Drenth 2011).

Table 9.1 New clinical trials

Drug	Sponsor	Enrollment (n)	Major inclusion criteria	Status	NCT Registry No.
Bosutinib	Pfizer	172	18–50 years, eGFR > 60 mL/min, TKV > 750 mL	Completed	NCT01233869
Tesevatinib (KD019)	Kadmon	120	≥22 years, eGFR ≥ 35 mL/min/1.73 m^2, htTKV ≥ 1000 mL/m	Recruiting	NCT01559363
Triptolide	Nanjing University	300	15–70 years, eGFR > 30 mL/min	Terminated due to drop outs	NCT00801268
Curcumin	Colorado University	68	6–25 years, eGFR > 80 mL/min	Recruiting	NCT02494141
Spironolactone	Colorado University	60	30–79 years, eGFR > 60 mL/min	Recruiting	NCT01853553

Abbreviations: *NCT* national clinical trial, *eGFR* estimated glomerular filtration rate, *TKV* total kidney volume, *htTKV* height-adjusted total kidney volume

9.3 Issues Related to Clinical Trials

9.3.1 Screening of High-Risk Group

In order to demonstrate the efficacy of candidate drugs within a limited period—for example, over three years, changes in outcome variables should be clearly shown. Outcome variables in ADPKD trials include kidney function, such as creatinine clearance, eGFR by the Cockcroft-Gault equation or MDRD formula, and time to ESRD, renal blood flow, structural changes, such as TKV, cyst volume, or parenchymal volume, and quality of life. Among these variables, kidney function and structural changes are most widely examined (Chapman 2009).

However, the trajectory of changes in those variables varies according to many factors. It is well known that kidney function remains stable for a long period, even with cyst growth and destruction of normal parenchyma. Once kidney function starts to deteriorate, ESRD progresses linearly but still takes more than 8 years in our experience. At the late stage of ADPKD, it is anticipated that anti-proliferative agents may be ineffective because a large portion of normal parenchyma has been replaced by cysts or pericystic fibrotic tissues. Therefore, only a small fraction of patients would be suitable for clinical trials. It is necessary to find the tools to identify the high-risk group who has near-normal kidney function currently but with a high probability of detectable changes within a short time period.

The risk factors for developing ESRD in ADPKD patients may include large TKV (Harris et al. 2006; Schrier et al. 2014b), PKD1 truncating mutation (Hateboer et al. 1999; Cornec-Le Gall et al. 2013), early diagnosis of ADPKD and early development of hypertension (Chapman et al. 2010a), albuminuria, and gross hematuria

(Chapman et al. 1994; Johnson and Gabow 1997). Among them, genotype is the most powerful predictor of ESRD. Patients with PKD1 truncating mutations showed a significantly higher risk of ESRD and death compared with patients with other genotypes, such as PKD1 in–frame insertion/deletion, PKD1 non-truncating mutations, or PKD2 mutations (Hwang et al. 2015). Along with the application of targeted exome sequencing in mutation detection in ADPKD, gene tests are expected to be more widely used in risk prediction (Choi et al. 2014). In the meantime, comprehensive acquisition of a family history of ESRD would be useful to predict the PKD1 genotype, for instance, family history of ESRD before the age of 55 (Barua et al. 2009; Pei 2011).

However, the severity or actual disease progression may vary even in patients from the same families. Radiologic evaluation to monitor structural changes would be useful to determine the time to intervene. TKV has been recognized as the most reliable structural parameter that can reflect a disease course from the early stages. When 241 subjects who participated in the CRISP study were followed over 8 years, the baseline height-adjusted TKV (htTKV) of more than 600 mL/m was found to be the most significant predictor among other factors such as age, serum creatinine, albuminuria, and MCP-1 (Chapman et al. 2012). Using CRISP and TEMPO data, Mayo researchers suggested the criteria for screening the high-risk group according to age and the percent annual change in htTKV. Subjects with typical ADPKD were classified into five groups as follows: 1A, the annual htTKV change <1.5 %; 1B, 1.5–3.0 %; 1C, 3.0–4.5 %; 1D, 4.5–6 %; and 1E, >6 % (Irazabal et al. 2015). It would be prudent to enroll subjects in the 1D or 1E classes in clinical trials (Fig. 9.2). Note that the htTKV on the y-axis is not evenly spaced, representing the exponential growth of TKV in Fig. 9.2. Future kidney function can be predicted by using an online calculator provided by Mayo researchers (http://www.mayo.edu/research/documents/pkd-center-adpkd-classification/doc-20094754) (Irazabal et al. 2015).

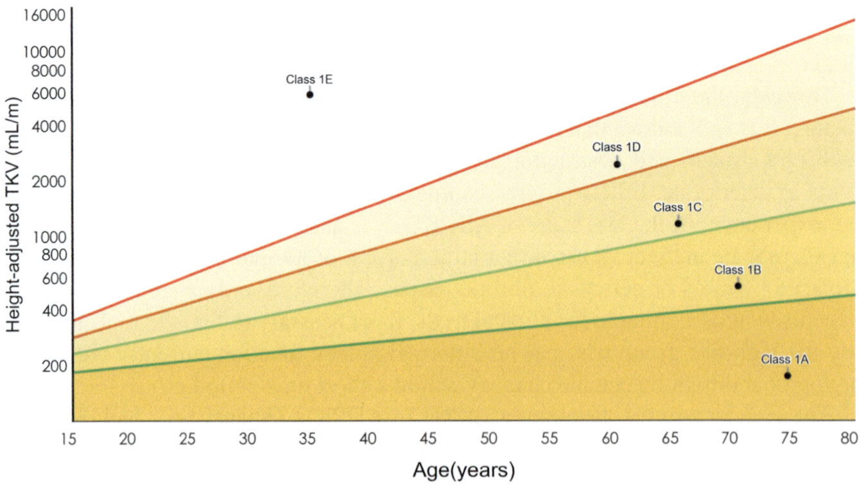

Fig. 9.2 Risk classification of ADPKD according to age and height-adjusted total kidney volume (htTKV)

Table 9.2 Suggested ideal candidates for clinical trials of ADPKD

Suggested ideal candidates for clinical trials of ADPKD
Protein-truncating mutation in PKD1 or family history of end-stage renal disease < 55 years
Height-adjusted total kidney volume (htTKV) > 600 mL/m or change in htTKV > 4.5 % per year
Estimated GFR > 45 mL/min/1.73 m^2

In summary, ideal candidates for clinical trials can be selected based on genotyping, radiologic evaluation, and/or renal function (Table 9.2).

9.3.2 *Assessment of Outcomes*

Traditionally, the hard outcome for clinical trials of kidney diseases is ESRD. Doubling of sCr or 50 % reduction of GFR is also considered to be a reliable surrogate marker. Recently, the U.S. FDA and National Kidney Foundation proposed alternative surrogate end points of a 30 % and 40 % decline in GFR from baseline instead of traditional end points (Rosansky and Glassock 2014). This suggestion is partly based on the observation that the renal function trajectory is variable in many patients with CKD (Li et al. 2012). Commentary by Rosansky et al. has suggested the change in renal function trajectory slope as a new surrogate marker (Rosansky and Glassock 2014). As seen in the ALADIN trial and in the TEMPO trial, the short-term acceleration of decline in GFR after intervention is not uncommon, and thus this finding should be incorporated in the design and analysis of drug efficacy trials.

In addition, which methods for evaluating GFR would be best is undetermined for ADPKD. Each trial uses different methods to measure and estimate GFR; for example, iohexol clearance in ALADIN and the reciprocal of sCr in the TEMPO study. This difference makes it difficult to compare the results in clinical trials. CRISP researchers who have compared methods for determining kidney function decline in early ADPKD reported that measured GFR was a stronger predictor than sCr-based estimation (Rule et al. 2006).

Another frequently used end point in ADPKD is a structural change over time. TKV can be measured by using ultrasound, CT, or MRI. Ellipsoid TKV or stereologic TKV is usually used to monitor treatment effects in clinical trials (Chapman et al. 2015). The ellipsoid estimation by CT or MRI is easy to use and strongly correlated with the stereologic method, and thus can be conveniently used in a clinical setting (Irazabal et al. 2015). Recently, fast and nearly automated techniques of kidney segmentation that automatically compute TKV have been reported. These methods of measuring TKV are expected to be available soon (Turco et al. 2015).

Various serum or urine biomarkers that have been suggested in studies on acute kidney injury and CKD are being evaluated in the field of ADPKD (Meijer et al. 2010). Particularly, the urine biomarkers that reflect the activity of the renal renin-angiotensin system or aberrant cellular signaling would be useful for short-term pilot studies or assessment of the drug mechanism of action (Park et al. 2015; Kistler et al. 2013). The issue of biomarkers in ADPKD was reviewed in detail in the previous chapter.

9.3.3 When to Intervene

It is still unclear when to start intervention for ADPKD. In case of V2RA, an early start of OPC-31260 in conditional, kidney epithelium-specific Pkd1-knockout mice was more effective than a late start in reducing cyst growth. Blunted physiologic effects were shown in late start group even with a high dose (0.1 %), compared to an early start with low dose (0.05 %) (Meijer et al. 2011). Given that the long-term adverse effects were minimal, it would be reasonable to initiate tdrug therapy from the early stage. Further analysis with clinical data according to the CKD stages would be informative when enough data have been accumulated.

9.4 Conclusion

With advances in the understanding of the pathogenesis of ADPKD, the number of drugs in the pipeline and number of clinical trials are both exponentially increasing. From the practical point of view, long-term safety and efficacy are the top priorities in the development of drugs because of the slow progressive nature of the disease. In addition, the development of experimental models that can mimic adult-onset PKD would facilitate drug discovery for ADPKD.

Acknowledgement The authors would like to express appreciation for the great help of Professors Curie Ahn (Seoul National University Hospital) and Kyubeck Lee (Kangbuk Samsung Hospital), who have devoted themselves to managing patients with ADPKD over the past two decades. We would also like to extend our gratitude to Danang Sarang (www.pkdkorea.co.kr/), a group for patients with ADPKD in Korea, and the PKD Foundation International (http://www.pkdinternational.org/).

References

Aguiari G, Catizone L, Del Senno L (2013) Multidrug therapy for polycystic kidney disease: a review and perspective. Am J Nephrol 37(2):175–182. doi:10.1159/000346812

Barash I, Ponda MP, Goldfarb DS, Skolnik EY (2010) A pilot clinical study to evaluate changes in urine osmolality and urine cAMP in response to acute and chronic water loading in autosomal dominant polycystic kidney disease. Clin J Am Soc Nephrol CJASN 5(4):693–697. doi:10.2215/CJN.04180609

Barua M, Cil O, Paterson AD, Wang K, He N, Dicks E, Parfrey P, Pei Y (2009) Family history of renal disease severity predicts the mutated gene in ADPKD. J Am Soc Nephrol JASN 20(8):1833–1838. doi:10.1681/ASN.2009020162

Bennett WM (2009) Autosomal dominant polycystic kidney disease: 2009 update for internists. Korean J Intern Med 24(3):165–168. doi:10.3904/kjim.2009.24.3.165

Berl T (2015) Vasopressin antagonists. N Engl J Med 372(23):2207–2216. doi:10.1056/NEJMra1403672

Bolignano D, Palmer SC, Ruospo M, Zoccali C, Craig JC, Strippoli GF (2015) Interventions for preventing the progression of autosomal dominant polycystic kidney disease. Cochrane Database Syst Rev 7:Cd010294. doi:10.1002/14651858.CD010294.pub2

Canaud G, Knebelmann B, Harris PC, Vrtovsnik F, Correas JM, Pallet N, Heyer CM, Letavernier E, Bienaime F, Thervet E, Martinez F, Terzi F, Legendre C (2010) Therapeutic mTOR inhibition in autosomal dominant polycystic kidney disease: what is the appropriate serum level? Am J Transplant Off J Am Soc Transplant Am Soc Transplant Surg 10(7):1701–1706. doi:10.1111/j.1600-6143.2010.03152.x

Caroli A, Perico N, Perna A, Antiga L, Brambilla P, Pisani A, Visciano B, Imbriaco M, Messa P, Cerutti R, Dugo M, Cancian L, Buongiorno E, De Pascalis A, Gaspari F, Carrara F, Rubis N, Prandini S, Remuzzi A, Remuzzi G, Ruggenenti P, ALADIN study group (2013) Effect of longacting somatostatin analogue on kidney and cyst growth in autosomal dominant polycystic kidney disease (ALADIN): a randomised, placebo-controlled, multicentre trial. Lancet 382(9903):1485–1495. doi:10.1016/S0140-6736(13)61407-5

Chang MY, Ong A (2013) New treatments for autosomal dominant polycystic kidney disease. Br J Clin Pharmacol 76(4):524–535

Chapman AB (2009) Improving clinical trial design for inquiries into the mechanisms of cyst growth in ADPKD. Kidney Int 75(2):139–141. doi:10.1038/ki.2008.596

Chapman AB, Johnson A, Gabow PA, Schrier RW (1990) The renin–angiotensin–aldosterone system and autosomal dominant polycystic kidney disease. N Engl J Med 323(16):1091–1096

Chapman AB, Johnson AM, Gabow PA, Schrier RW (1994) Overt proteinuria and microalbuminuria in autosomal dominant polycystic kidney disease. J Am Soc Nephrol JASN 5(6):1349–1354

Chapman AB, Stepniakowski K, Rahbari-Oskoui F (2010a) Hypertension in autosomal dominant polycystic kidney disease. Adv Chronic Kidney Dis 17(2):153–163. doi:10.1053/j.ackd.2010.01.001

Chapman AB, Torres VE, Perrone RD, Steinman TI, Bae KT, Miller JP, Miskulin DC, Rahbari Oskoui F, Masoumi A, Hogan MC, Winklhofer FT, Braun W, Thompson PA, Meyers CM, Kelleher C, Schrier RW (2010b) The HALT polycystic kidney disease trials: design and implementation. Clin J Am Soc Nephrol CJASN 5(1):102–109. doi:10.2215/CJN.04310709

Chapman AB, Bost JE, Torres VE, Guay-Woodford L, Bae KT, Landsittel D, Li J, King BF, Martin D, Wetzel LH, Lockhart ME, Harris PC, Moxey-Mims M, Flessner M, Bennett WM, Grantham JJ (2012) Kidney volume and functional outcomes in autosomal dominant polycystic kidney disease. Clin J Am Soc Nephrol CJASN 7(3):479–486. doi:10.2215/CJN.09500911

Chapman AB, Devuyst O, Eckardt KU, Gansevoort RT, Harris T, Horie S, Kasiske BL, Odland D, Pei Y, Perrone RD, Pirson Y, Schrier RW, Torra R, Torres VE, Watnick T, Wheeler DC, Conference Participants (2015) Autosomal-Dominant Polycystic Kidney Disease (ADPKD): executive summary from a Kidney Disease: Improving Global Outcomes (KDIGO) controversies conference. Kidney Int. doi:10.1038/ki.2015.59

Chen D, Ma Y, Wang X, Yu S, Li L, Dai B, Mao Z, Liu H, Liu S, Mei C (2014) Triptolide-containing formulation in patients with autosomal dominant polycystic kidney disease and proteinuria: an uncontrolled trial. Am J Kidney Dis Off J Nat Kidney Found 63(6):1070

Choi R, Park HC, Lee K, Lee MG, Kim JW, Ki CS, Hwang YH, Ahn C (2014) Identification of novel PKD1 and PKD2 mutations in Korean patients with autosomal dominant polycystic kidney disease. BMC Med Genet 15(1):129. doi:10.1186/s12881-014-0129-y

Chrispijn M, Drenth JP (2011) Everolimus and long acting octreotide as a volume reducing treatment of polycystic livers (ELATE): study protocol for a randomized controlled trial. Trials 12:246. doi:10.1186/1745-6215-12-246

Cornec-Le Gall E, Audrezet MP, Chen JM, Hourmant M, Morin MP, Perrichot R, Charasse C, Whebe B, Renaudineau E, Jousset P, Guillodo MP, Grall-Jezequel A, Saliou P, Ferec C, Le Meur Y (2013) Type of PKD1 mutation influences renal outcome in ADPKD. J Am Soc Nephrol JASN 24(6):1006–1013. doi:10.1681/ASN.2012070650

Ecder T, Edelstein CL, Fick-Brosnahan GM, Johnson AM, Chapman AB, Gabow PA, Schrier RW (2001) Diuretics versus angiotensin-converting enzyme inhibitors in autosomal dominant polycystic kidney disease. Am J Nephrol 21(2):98–103

Elliott J, Zheleznova NN, Wilson PD (2011) c-Src inactivation reduces renal epithelial cell-matrix adhesion, proliferation, and cyst formation. Am J Phys Cell Phys 301(2):C522–C529

Gansevoort RT, Arici M, Benzing T, Birn H, Capasso G, Covic A, Devuyst O, Drechsler C, Eckardt KU, Emma F, Knebelmann B, Le Meur Y, Massy ZA, Ong AC, Ortiz A, Schaefer F, Torra R, Vanholder R, Wiecek A, Zoccali C, Van Biesen W (2016) Recommendations for the use of tolvaptan in autosomal dominant polycystic kidney disease: a position statement on behalf of the ERA-EDTA working groups on inherited kidney disorders and European renal best practice. Nephrol Dial Trans Off Publ Eur Dial Trans Assoc Eur Renal Assoc 31(3):337–348. doi:10.1093/ndt/gfv456

Gao J, Zhou H, Lei T, Zhou L, Li W, Li X, Yang B (2011) Curcumin inhibits renal cyst formation and enlargement in vitro by regulating intracellular signaling pathways. Eur J Pharmacol 654(1):92–99

Gevers TJ, Chrispijn M, Wetzels JF, Drenth JP (2012) Rationale and design of the RESOLVE trial: lanreotide as a volume reducing treatment for polycystic livers in patients with autosomal dominant polycystic kidney disease. BMC Nephrol 13:17. doi:10.1186/1471-2369-13-17

Harris PC, Bae KT, Rossetti S, Torres VE, Grantham JJ, Chapman AB, Guay-Woodford LM, King BF, Wetzel LH, Baumgarten DA, Kenney PJ, Consugar M, Klahr S, Bennett WM, Meyers CM, Zhang QJ, Thompson PA, Zhu F, Miller JP (2006) Cyst number but not the rate of cystic growth is associated with the mutated gene in autosomal dominant polycystic kidney disease. J Am Soc Nephrol JASN 17(11):3013–3019. doi:10.1681/ASN.2006080835

Hateboer N, v Dijk MA, Bogdanova N, Coto E, Saggar-Malik AK, San Millan JL, Torra R, Breuning M, Ravine D (1999) Comparison of phenotypes of polycystic kidney disease types 1 and 2. European PKD1-PKD2 study group. Lancet 353(9147):103–107

Higashihara E, Nutahara K, Tanbo M, Hara H, Miyazaki I, Kobayashi K, Nitatori T (2014) Does increased water intake prevent disease progression in autosomal dominant polycystic kidney disease? Nephrol Dial Trans Off Publ Eur Dial Trans Assoc Eur Renal Assoc 29(9):1710–1719. doi:10.1093/ndt/gfu093

Hopp K, Hommerding CJ, Wang X, Ye H, Harris PC, Torres VE (2015) Tolvaptan plus pasireotide shows enhanced efficacy in a PKD1 model. J Am Soc Nephrol JASN 26(1):39–47. doi:10.1681/ASN.2013121312

Hwang YH, Conklin J, Chan W, Roslin NM, Liu J, He N, Wang K, Sundsbak JL, Heyer CM, Haider M, Paterson AD, Harris PC, Pei Y (2015) Refining genotype-phenotype correlation in autosomal dominant polycystic kidney disease. J Am Soc Nephrol JASN. doi:10.1681/asn.2015060648

Irazabal MV, Torres VE (2013) Experimental therapies and ongoing clinical trials to slow down progression of ADPKD. Curr Hypertens Rev 9(1):44

Irazabal MV, Rangel LJ, Bergstralh EJ, Osborn SL, Harmon AJ, Sundsbak JL, Bae KT, Chapman AB, Grantham JJ, Mrug M, Hogan MC, El-Zoghby ZM, Harris PC, Erickson BJ, King BF, Torres VE, Investigators C (2015) Imaging classification of autosomal dominant polycystic kidney disease: a simple model for selecting patients for clinical trials. J Am Soc Nephrol JASN 26(1):160–172. doi:10.1681/ASN.2013101138

Johnson AM, Gabow PA (1997) Identification of patients with autosomal dominant polycystic kidney disease at highest risk for end-stage renal disease. J Am Soc Nephrol JASN 8(10):1560–1567

Kistler AD, Serra AL, Siwy J, Poster D, Krauer F, Torres VE, Mrug M, Grantham JJ, Bae KT, Bost JE, Mullen W, Wuthrich RP, Mischak H, Chapman AB (2013) Urinary proteomic biomarkers for diagnosis and risk stratification of autosomal dominant polycystic kidney disease: a multicentric study. PLoS One 8(1):e53016. doi:10.1371/journal.pone.0053016

Leonhard WN, van der Wal A, Novalic Z, Kunnen SJ, Gansevoort RT, Breuning MH, de Heer E, Peters DJ (2011) Curcumin inhibits cystogenesis by simultaneous interference of multiple signaling pathways: in vivo evidence from a Pkd1-deletion model. Am J Physiol Renal Physiol 300(5):F1193–F1202

Leuenroth SJ, Bencivenga N, Chahboune H, Hyder F, Crews CM (2010) Triptolide reduces cyst formation in a neonatal to adult transition Pkd1 model of ADPKD. Nephrol Dial Trans Off Publ Eur Dial Trans Assoc Eur Renal Assoc 25(7):2187–2194. doi:10.1093/ndt/gfp777

Li L, Astor BC, Lewis J, Hu B, Appel LJ, Lipkowitz MS, Toto RD, Wang X, Wright JT Jr, Greene TH (2012) Longitudinal progression trajectory of GFR among patients with CKD. Am J Kidney Dis Off J Nat Kidney Found 59(4):504–512. doi:10.1053/j.ajkd.2011.12.009

Meijer E, Boertien WE, Nauta FL, Bakker SJ, van Oeveren W, Rook M, van der Jagt EJ, van Goor H, Peters DJ, Navis G, de Jong PE, Gansevoort RT (2010) Association of urinary biomarkers with disease severity in patients with autosomal dominant polycystic kidney disease: a cross-sectional analysis. Am J Kidney Dis Off J Nat Kidney Found 56(5):883–895. doi:10.1053/j.ajkd.2010.06.023

Meijer E, Gansevoort RT, de Jong PE, van der Wal AM, Leonhard WN, de Krey SR, van den Born J, Mulder GM, van Goor H, Struck J, de Heer E, Peters DJ (2011) Therapeutic potential of vasopressin V2 receptor antagonist in a mouse model for autosomal dominant polycystic kidney disease: optimal timing and dosing of the drug. Nephrol Dial Trans Off Publ Eur Dial Trans Assoc Eur Renal Assoc 26(8):2445–2453. doi:10.1093/ndt/gfr069

Meijer E, Drenth JP, d'Agnolo H, Casteleijn NF, de Fijter JW, Gevers TJ, Kappert P, Peters DJ, Salih M, Soonawala D, Spithoven EM, Torres VE, Visser FW, Wetzels JF, Zietse R, Gansevoort RT, Consortium D (2014) Rationale and design of the DIPAK 1 study: a randomized controlled clinical trial assessing the efficacy of lanreotide to Halt disease progression in autosomal dominant polycystic kidney disease. Am J Kidney Dis Off J Nat Kidney Found 63(3):446–455. doi:10.1053/j.ajkd.2013.10.011

Muto S, Kawano H, Higashihara E, Narita I, Ubara Y, Matsuzaki T, Ouyang J, Torres VE, Horie S (2015) The effect of tolvaptan on autosomal dominant polycystic kidney disease patients: a subgroup analysis of the Japanese patient subset from TEMPO 3:4 trial. Clin Exp Nephrol. doi:10.1007/s10157-015-1086-2

Nagao S, Nishii K, Katsuyama M, Kurahashi H, Marunouchi T, Takahashi H, Wallace DP (2006) Increased water intake decreases progression of polycystic kidney disease in the PCK rat. J Am Soc Nephrol JASN 17(8):2220–2227. doi:10.1681/asn.2006030251

Nagao S, Nishii K, Yoshihara D, Kurahashi H, Nagaoka K, Yamashita T, Takahashi H, Yamaguchi T, Calvet JP, Wallace DP (2008) Calcium channel inhibition accelerates polycystic kidney disease progression in the Cy/+ rat. Kidney Int 73(3):269–277. doi:10.1038/sj.ki.5002629

Nielsen FS, Rossing P, Gall M-A, Skøtt P, Smidt UM, Parving H-H (1997) Long-term effect of lisinopril and atenolol on kidney function in hypertensive NIDDM subjects with diabetic nephropathy. Diabetes 46(7):1182–1188

Novalic Z, van der Wal AM, Leonhard WN, Koehl G, Breuning MH, Geissler EK, de Heer E, Peters DJ (2012) Dose-dependent effects of sirolimus on mTOR signaling and polycystic kidney disease. J Am Soc Nephrol JASN 23(5):842–853. doi:10.1681/ASN.2011040340

Park HC, Kang AY, Jang JY, Kim H, Han M, Oh KH, Kim SH, Noh JW, Cheong HI, Hwang YH, Ahn C (2015) Increased urinary Angiotensinogen/Creatinine (AGT/Cr) ratio may be associated with reduced renal function in autosomal dominant polycystic kidney disease patients. BMC Nephrol 16:86. doi:10.1186/s12882-015-0075-8

Pei Y (2011) Practical genetics for autosomal dominant polycystic kidney disease. Nephron Clin Pract 118(1):c19–c30. doi:10.1159/000320887

Redon J, Mancia G, Sleight P, Schumacher H, Gao P, Pogue J, Fagard R, Verdecchia P, Weber M, Bohm M, Williams B, Yusoff K, Teo K, Yusuf S (2012) Safety and efficacy of low blood pressures among patients with diabetes: subgroup analyses from the ONTARGET (ONgoing Telmisartan Alone and in combination with Ramipril Global Endpoint Trial). J Am Coll Cardiol 59(1):74–83. doi:10.1016/j.jacc.2011.09.040

Rosansky SJ, Glassock RJ (2014) Is a decline in estimated GFR an appropriate surrogate end point for renoprotection drug trials? Kidney Int 85(4):723–727. doi:10.1038/ki.2013.506

Ruggenenti P, Remuzzi A, Ondei P, Fasolini G, Antiga L, Ene-Iordache B, Remuzzi G, Epstein FH (2005) Safety and efficacy of long-acting somatostatin treatment in autosomal-dominant polycystic kidney disease. Kidney Int 68(1):206–216. doi:10.1111/j.1523-1755.2005.00395.x

Rule AD, Torres VE, Chapman AB, Grantham JJ, Guay-Woodford LM, Bae KT, Klahr S, Bennett WM, Meyers CM, Thompson PA, Miller JP (2006) Comparison of methods for determining renal function decline in early autosomal dominant polycystic kidney disease: the consortium of radiologic imaging studies of polycystic kidney disease cohort. J Am Soc Nephrol JASN 17(3):854–862. doi:10.1681/asn.2005070697

Sarnak MJ, Greene T, Wang X, Beck G, Kusek JW, Collins AJ, Levey AS (2005) The effect of a lower target blood pressure on the progression of kidney disease: long-term follow-up of the modification of diet in renal disease study. Ann Intern Med 142(5):342–351

Schrier R, McFann K, Johnson A, Chapman A, Edelstein C, Brosnahan G, Ecder T, Tison L (2002) Cardiac and renal effects of standard versus rigorous blood pressure control in autosomal-dominant polycystic kidney disease: results of a seven-year prospective randomized study. J Am Soc Nephrol JASN 13(7):1733–1739

Schrier RW, Gross P, Gheorghiade M, Berl T, Verbalis JG, Czerwiec FS, Orlandi C (2006) Tolvaptan, a selective oral vasopressin V2-receptor antagonist, for hyponatremia. N Engl J Med 355(20):2099–2112. doi:10.1056/NEJMoa065181

Schrier RW, Abebe KZ, Perrone RD, Torres VE, Braun WE, Steinman TI, Winklhofer FT, Brosnahan G, Czarnecki PG, Hogan MC, Miskulin DC, Rahbari-Oskoui FF, Grantham JJ, Harris PC, Flessner MF, Bae KT, Moore CG, Chapman AB (2014a) Blood pressure in early autosomal dominant polycystic kidney disease. N Engl J Med 371(24):2255–2266. doi:10.1056/NEJMoa1402685

Schrier RW, Brosnahan G, Cadnapaphornchai MA, Chonchol M, Friend K, Gitomer B, Rossetti S (2014b) Predictors of autosomal dominant polycystic kidney disease progression. J Am Soc Nephrol JASN 25(11):2399–2418. doi:10.1681/asn.2013111184

Serra AL, Poster D, Kistler AD, Krauer F, Raina S, Young J, Rentsch KM, Spanaus KS, Senn O, Kristanto P, Scheffel H, Weishaupt D, Wuthrich RP (2010) Sirolimus and kidney growth in autosomal dominant polycystic kidney disease. N Engl J Med 363(9):820–829. doi:10.1056/NEJMoa0907419

Shillingford JM, Leamon CP, Vlahov IR, Weimbs T (2012) Folate-conjugated rapamycin slows progression of polycystic kidney disease. J Am Soc Nephrol JASN 23(10):1674–1681. doi:10.1681/ASN.2012040367

Sweeney WE, von Vigier RO, Frost P, Avner ED (2008) Src inhibition ameliorates polycystic kidney disease. J Am Soc Nephrol 19(7):1331–1341

Tao Y, Kim J, Schrier RW, Edelstein CL (2005) Rapamycin markedly slows disease progression in a rat model of polycystic kidney disease. J Am Soc Nephrol JASN 16(1):46–51. doi:10.1681/ASN.2004080660

Torres VE (2005) Vasopressin antagonists in polycystic kidney disease. Kidney Int 68(5):2405–2418

Torres VE, Bankir L, Grantham JJ (2009) A case for water in the treatment of polycystic kidney disease. Clin J Am Soc Nephrol CJASN 4(6):1140–1150. doi:10.2215/CJN.00790209

Torres VE, Meijer E, Bae KT, Chapman AB, Devuyst O, Gansevoort RT, Grantham JJ, Higashihara E, Perrone RD, Krasa HB, Ouyang JJ, Czerwiec FS (2011) Rationale and design of the TEMPO

(Tolvaptan Efficacy and Safety in Management of Autosomal Dominant Polycystic Kidney Disease and its Outcomes) 3–4 study. Am J Kidney Dis Off J Nat Kidney Found 57(5):692–699. doi:10.1053/j.ajkd.2010.11.029

Torres VE, Chapman AB, Devuyst O, Gansevoort RT, Grantham JJ, Higashihara E, Perrone RD, Krasa HB, Ouyang J, Czerwiec FS, Investigators TT (2012) Tolvaptan in patients with autosomal dominant polycystic kidney disease. N Engl J Med 367(25):2407–2418. doi:10.1056/NEJMoa1205511

Torres VE, Abebe KZ, Chapman AB, Schrier RW, Braun WE, Steinman TI, Winklhofer FT, Brosnahan G, Czarnecki PG, Hogan MC, Miskulin DC, Rahbari-Oskoui FF, Grantham JJ, Harris PC, Flessner MF, Moore CG, Perrone RD, Investigators H-PT (2014) Angiotensin blockade in late autosomal dominant polycystic kidney disease. N Engl J Med 371(24):2267–2276. doi:10.1056/NEJMoa1402686

Torres VE, Higashihara E, Devuyst O, Chapman AB, Gansevoort RT, Grantham JJ, Perrone RD, Ouyang J, Blais JD, Czerwiec FS (2016) Effect of tolvaptan in autosomal dominant polycystic kidney disease by CKD stage: results from the TEMPO 3:4 trial. Clin J Am Soc Nephrol CJASN. doi:10.2215/cjn.06300615

Turco D, Severi S, Mignani R, Magistroni R, Corsi C (2015) Geometry-independent assessment of renal volume in polycystic kidney disease from magnetic resonance imaging. Conf Proceed Annu Int Conf IEEE Eng Med Biol Soc IEEE Eng Med Biol Soc Annu Conf 2015:3081–3084. doi:10.1109/embc.2015.7319043

van Keimpema L, Nevens F, Vanslembrouck R, van Oijen MG, Hoffmann AL, Dekker HM, de Man RA, Drenth JP (2009) Lanreotide reduces the volume of polycystic liver: a randomized, double-blind, placebo-controlled trial. Gastroenterology 137(5):1661–1662. e1661–e1668. doi:10.1053/j.gastro.2009.07.052

Wahl PR, Serra AL, Le Hir M, Molle KD, Hall MN, Wuthrich RP (2006) Inhibition of mTOR with sirolimus slows disease progression in Han:SPRD rats with autosomal dominant polycystic kidney disease (ADPKD). Nephrol Dial Trans Off Publ Eur Dial Trans Assoc Eur Renal Assoc 21(3):598–604. doi:10.1093/ndt/gfi181

Walz G, Budde K, Mannaa M, Nurnberger J, Wanner C, Sommerer C, Kunzendorf U, Banas B, Horl WH, Obermuller N, Arns W, Pavenstadt H, Gaedeke J, Buchert M, May C, Gschaidmeier H, Kramer S, Eckardt KU (2010) Everolimus in patients with autosomal dominant polycystic kidney disease. N Engl J Med 363(9):830–840. doi:10.1056/NEJMoa1003491

Wang X, Wu Y, Ward CJ, Harris PC, Torres VE (2008) Vasopressin directly regulates cyst growth in polycystic kidney disease. J Am Soc Nephrol JASN 19(1):102–108. doi:10.1681/ASN.2007060688

Watkins PB, Lewis JH, Kaplowitz N, Alpers DH, Blais JD, Smotzer DM, Krasa H, Ouyang J, Torres VE, Czerwiec FS, Zimmer CA (2015) Clinical pattern of tolvaptan-associated liver injury in subjects with autosomal dominant polycystic kidney disease: analysis of clinical trials database. Drug Saf 38(11):1103–1113. doi:10.1007/s40264-015-0327-3

Watnick T, Germino GG (2010) mTOR inhibitors in polycystic kidney disease. N Engl J Med 363(9):879–881. doi:10.1056/NEJMe1006925

Yusuf S, Teo KK, Pogue J, Dyal L, Copland I, Schumacher H, Dagenais G, Sleight P, Anderson C (2008) Telmisartan, ramipril, or both in patients at high risk for vascular events. N Engl J Med 358(15):1547–1559. doi:10.1056/NEJMoa0801317

Index

A
ACEI/ARB, 108
ACEIs. *See* Angiotensin-converting enzyme inhibitors (ACEIs)
Adenylyl cyclase 6, 27
Akt, 26
Albuminuria, 90
Allelic heterogeneity, 87
AMP-activated protein kinase (AMPK), 50
AMPK activator, 78
Angiotensin converting enzyme (ACE), 88
Angiotensin-converting enzyme inhibitors (ACEIs), 78, 107
Angiotensin receptor blockers (ARBs), 107
AntagomiR, 65
Anterograde transport, 47
AntimiR, 65
Apelin, 95
Apelin-APJ axis, 95
Apoptosis, 29
ARBs. *See* Angiotensin receptor blockers (ARBs)
Arginine vasopressin (AVP), 96
Asymmetric dimethylarginine (ADMA), 94
AVP. *See* Arginine vasopressin (AVP)
AVP receptor blockers, 106
5-Aza-2′-deoxycytidine, 60, 63

B
Basal bodies, 47
Bax, 30
Bcl-2, 30
Biomarker, 86
Bosutinib, 6
B-Raf, 26

C
Calcium, 49
Calcium channels, 51
Calcium-induced calcium release (CICR), 53
cAMP. *See* Cyclic adenosine monophosphate (cAMP)
cAMP response element binding protein (CREB), 27
Cardiac defects, 4
Caspases, 29
β-Catenin dependent, 16
β-Catenin independent, 16
CDCA, 53
CDKs. *See* Cyclin-dependent kinases (CDKs)
CDK5, 31
Cell cycle, 25
Cell proliferation, 25
Chromatin, 59
Chronic kidney disease (CKD), 108
CICR. *See* Calcium-induced calcium release (CICR)
Cilia, 45
Ciliogenesis, 46
Ciliopathies, 51
CKD. *See* Chronic kidney disease (CKD)
c-Myc, 30
Computed tomography (CT), 93
Congenital polycystic kidney (cpk), 73
Consortium for Radiologic Imaging Study of PKD (CRISP), 87
Copeptin, 96
CpG islands, 60
Cpk. *See* Congenital polycystic kidney (cpk)
CREB. *See* cAMP response element binding protein (CREB)
Curcumin, 112

J.H. Park, C. Ahn (eds.), *Cystogenesis*, Advances in Experimental Medicine and Biology 933, DOI 10.1007/978-981-10-2041-4

Cyclic adenosine monophosphate (cAMP), 5, 15, 27
Cyclin-dependent kinases (CDKs), 26
Cys1 gene, 73
Cyst rupture, 90
Cytochrome c, 29
Cytosine-guanosine dinucleotides (CpGs), 60

D
de novo methylation, 60
de novo mutation, 4
Death receptors, 29
Development, 50
Dialysis, 4
Dicer, 61
Dickkopf 3 (DKK3), 88
DNA methylation, 7, 60, 61
DNA methyltransferases (DNMTs), 60
Drosha, 61
Dynein-2, 47

E
4E-binding protein 1, 28
EGF. *See* Epidermal growth factor (EGF)
eIF2, 28
Endothelial nitric oxide synthase, 88
Epidermal growth factor (EGF), 27
Epigenetic modification, 88
Epigenetic regulation, 59
ESRD, 4
Everolimus, 6, 76
Exportin-5, 61
Extracellular matrix, 7
Extracellular signal-regulated kinase (ERK), 26

F
Fibrosis, 36
Filtration, 7
Folate-conjugated drugs, 77
Food restriction, 79

G
G-protein coupled receptors (GPCRs), 30
Gene-body methylation, 60
Genome-wide DNA methylation, 62
Germline mutation, 4
Glioma (Gli), 48
Glomerular filtration rate(GFR), 91
Gross hematuria, 90
Growth factors, 5

H
Hairpin structure, 61
HALT-PKD study, 107
Han:SPRD Cy rat model, 74
Han:SPRD rat, 29
HDAC inhibitor, 64
Hedgehog, 48
Hematuria, 90
Hepatocystic disease, 4
High water intake, 106
Histone, 59
Histone deacetylases (HDACs), 60
Histone methyltransferases (HMTs), 60
Histone modification, 60
HMG-CoA reductase, 77
Homeostasis, 7
Homeostatic, 5
Hypermethylation, 60, 62
Hypertension, 4, 90, 107
Hypomorphic alleles, 88

I
IAPs. *See* Inhibitor of apoptosis proteins (IAPs)
Ift20-targeted mouse model, 73
Inflammation, 35
Inhibitor of apoptosis proteins (IAPs), 29
Inositol trisphosphate receptor (IP3R), 26
Integrins, 7
Intracranial aneurysms, 4
Intraflagellar transport, 47

J
JAK-STAT, 16, 38
Janus kinase, 28
Juvenile cystic kidney (jck), 52, 73

K
KD019, 112
Kif3a, 53
Kinesin-II, 47

L
Laminins, 7
Left ventricular mass index (LVMI), 108

Linkage analysis, 87
Lkb1, 50
Locus heterogeneity, 87
Lovastin, 79
Luminal flow, 49
LVMI. *See* Left ventricular mass index (LVMI)

M
Macrophage, 36
Magnetic resonance imaging (MRI), 93
Mammalian target of rapamycin (mTOR) inhibitors, 76–77
Manifestations, 4
MAPKs. *See* Mitogen-activated protein kinases (MAPKs)
Matrix metalloproteinases (MMPs), 7, 39
Mature miRNA, 61
MDRD, 107
Mechanosensors, 50
Mechanosignaling, 49
mechanotransduction, 50
MEK, 26
MEK inhibitors, 77
Methylated cytosine (5-methylcytosine), 60
Microalbuminuria, 91
MicroRNA, 8, 61
Mitogen-activated protein kinases (MAPKs), 26
MMPs. *See* Matrix metalloproteinases (MMPs)
Modification of Diet in Renal Disease (MDRD), 91
Modifier gene effect, 88
Monocyte chemotactic protein-1 (MCP-1), 6, 36, 37, 92, 94, 114
Motile cilia, 46
mTOR, 16, 26, 50
mTOR signaling, 6
mTORC1, 50
mTORC2, 50

N
N-acetyl-β-D glucosaminidase (NAG), 94
Nek8 gene, 52, 73
Neutrophil gelatinase-associated lipocalin (NGAL), 94
NF-kB, 37
Nuclear RNase III enzyme, 61

O
Oak ridge polycystic kidney (ORPK) mouse, 72
Octreotide, 75, 111

P
p21Waf1, 28
P53, 30
PC1, 13
PC2, 14
PERK, 28
Phosphatidylinositide 3-kinases, 28
PKD rodent models, 72
PKD1, 13
Pkd1-targeted mice, 72
PKD2, 13
Pkd2 constitutive knockout mice, 72
PKD2 transgenic mice, 72
PKHD1, 4
Polymorphism, 88
Pravastatin, 77
Pre-miRNA, 61
Primary cilia, 45
Pri-miRNA, 61
PRKCSH, 3
Proliferation, 50
Protein kinase B, 28
Proteinuria, 90
Ptch1, 48
Pyrrolidine dithiocarbamate, 78
Pyuria, 91

R
Raf inhibitor, 77
Rapamycin, 6, 76, 109
RAS blockers, 109
Renin-angiotensin system (RAS), 28, 90
Retrograde transport, 47
Rheb, 28
RNA polymerase II, 61
RNA-induced silencing complex (RISC), 61
Roscovitine, 79

S
S-adenosyl methionine, 60
SEC63, 3
Second hit, 4, 17

Secreted Frizzled-related protein 4 (sFRP4), 97
Serum creatinine, 91
Shear stress, 50
Signal transducer and activator of transcription, 28
Signaling pathways, 15
Sirolimus, 6, 76, 109
Smo, 48
Somatostatin analogs, 30, 75–76
Sorafenib, 6
Spironolactone, 112
STAT3, 6
Statins, 77
Stem-loop structure, 61
Steviol, 78
SUISSE study, 109
Syndecans, 7

T
TEMPO 3:4 trial, 75
Tesevatinib, 112
TGF-β, 38
Third hit, 4
TIMP, 39
Tissue inhibitors of metalloproteinases (TIMPs), 7
TNF-α, 37
Tolvaptan, 75
Transplantation, 4
Trichostatin A, 64
Triptolide, 112
TRPC1, 27
TRPV4, 27
TSC1/2, 27
α-Tubulin, 64
Tumor suppressor genes, 60
α1 Type IV collagen, 7

U
Ultrasonography, 93
3′-Untranslated region, 61
Urinary angiotensinogen, 95
Urinary tract infection(UTI), 91

V
Vasopressin, 27
Vasopressin V2 receptor (V2R), 6, 27
Vasopressin V2 receptor antagonist (V2RAs), 30, 75, 110

W
Wnt, 39, 52
Wnt signaling pathways, 16

Z
Zebularine, 63

MIX
Papier aus verantwortungsvollen Quellen
Paper from responsible sources
FSC® C105338

If you have any concerns about our products,
you can contact us on
ProductSafety@springernature.com

In case Publisher is established outside the EU,
the EU authorized representative is:
Springer Nature Customer Service Center GmbH
Europaplatz 3, 69115 Heidelberg, Germany

Printed by Libri Plureos GmbH
in Hamburg, Germany